Exclusion in Smart Cities

Smart Cities are fascinating, but they also have a dark side that little is said or written about. One issue is the possibility of generating various types of exclusion. For this reason, this book will develop principles for assessing the inclusiveness of Smart Cities and outline strategies for strengthening urban resistance to exclusion.

The book will consider the essence, scale and types of exclusion and include analysis of inclusiveness in practice in the assessment and activities of Smart Cities. The research assesses Smart City literature, with particular emphasis on criticism of the Smart City concept and exclusion generated in smart urban structures, critically analyses the principles and indicators used in international inclusive Smart City rankings, highlights case studies on best practices for preventing exclusions in Smart Cities, and uses trend analysis to assess the scale and intensity of exclusion threats.

This book can be used as a compendium of knowledge about exclusions in Smart Cities by both researchers and students, as well as all urban stakeholders, in particular city decision-makers.

Izabela Jonek-Kowalska is Full Professor in the Department of Economics and Computer Sciences at Silesian University of Technology, Poland. For over 23 years she has been dealing with the economics of business entities, including finance and management of Smart Cities. Her research output includes over 400 publications with national and international publishers, as well as many undertakings and projects implemented in cooperation with the business environment. Izabela's research interests include finance of business entities, Smart City risk management, and municipal and industrial economics.

Radosław Wolniak is Full Professor in the Department of Economics and Informatics at Silesian University of Technology, Poland. For over 24 years he has been dealing with the issues of management, including Industry 4.0, Smart City, digitalization, and artificial intelligence. His scientific achievements include over 600 publications with national and international publishers, as well as many undertakings and projects implemented in cooperation with the business environment. Radosław's research interests include management, Smart Cities, Industry 4.0, digitalization, innovativeness, and artificial intelligence.

Routledge Research in Planning and Urban Design

Routledge Research in Planning and Urban Design is a series of academic monographs for scholars working in these disciplines and the overlaps between them. Building on Routledge's history of academic rigour and cutting-edge research, the series contributes to the rapidly expanding literature in all areas of planning and urban design.

Contested Airport Land
Social-Spatial Transformation and Environmental Injustice in Asia and Africa
Edited by Irit Ittner, Sneha Sharma, Isaac Khambule and Hanna Geschewski

Rebuilding Urban Complexity
A Configurational Approach to Postindustrial Cities
Francesca Froy

The Protection of Green Spaces for Climate Change Adaptation
Planning Systems, Policies and Instruments
Edited by Maciej J. Nowak

Exclusion in Smart Cities
Assessment and Strategy for Strengthening Resilience
Izabela Jonek-Kowalska and Radosław Wolniak

For more information about this series, please visit: www.routledge.com/Routledge-Research-in-Planning-and-Urban-Design/book-series/RRPUD

Exclusion in Smart Cities

Assessment and Strategy for Strengthening Resilience

Izabela Jonek-Kowalska and Radosław Wolniak

Designed cover image: © Hiroshi Watanabe / Getty Images

First published 2025
by Routledge
4 Park Square, Milton Park, Abingdon, Oxon OX14 4RN

and by Routledge
605 Third Avenue, New York, NY 10017

Routledge is an imprint of the Taylor & Francis Group, an informa business

British Library Cataloguing-in-Publication Data
A catalogue record for this book is available from the British Library

ISBN: 978-1-032-81460-5 (hbk)
ISBN: 978-1-032-81465-0 (pbk)
ISBN: 978-1-003-49999-2 (ebk)

DOI: 10.4324/9781003499992

The Open Access version of Chapter 1 was funded by Silesian University of Technology.

Contents

Figures

Tables

Introduction

The concept of Smart City is developing very dynamically in response to the search for ways to improve the quality of life in modern societies. It is worth emphasizing that this is a development not only referred to in the pages of scientific literature but also reflected in the actual strategies and activities of many cities around the world. The mass nature of Smart City solutions prompts us to assess their effectiveness and usefulness. It also provides numerous reasons for less visible but necessary and creative criticism of the implementation of the Smart City concept. For this reason, the authors of this monograph are looking for answers to the following research problems:

1 What is the scope and scale of threats related to various dimensions of exclusion that Smart Cities may generate?
2 How to assess the maturity level of Smart City inclusiveness?
3 How can different dimensions of exclusion be prevented in Smart Cities?

In the context of the above research problems, the main goal of the research was to develop principles for assessing the inclusiveness of Smart Cities and strategies for strengthening urban resistance to exclusion. The following detailed objectives subordinated to this goal have been implemented in the subsequent chapters of this monograph:

- to present the evolution of the concept of Smart Cities,
- to identify the reasons for criticism of the Smart City concept in theory and practice,

DOI: 10.4324/9781003499992-1

- to conduct research on the dimensions of exclusion of Smart Cities using literature review, critical analysis of international Smart Cities, case studies, and statistical data from an international perspective,
- to design a methodology for assessing the inclusiveness of a Smart City,
- to develop guidelines for a strategy strengthening the resilience of Smart Cities to various dimensions of exclusion,
- to identify the scale and intensity of future exclusion threats based on the analysis of statistical data.

The monograph uses the following **research methods**:

1 literature studies in the area of Smart City, with particular emphasis on criticism of the Smart City concept and exclusion generated in smart urban structures;
2 critical analysis of the principles and indicators used in international, inclusive Smart City rankings;
3 case studies on best practices for preventing exclusions in Smart City;
4 trend analysis aimed at assessing the scale and intensity of exclusion threats carried out on the basis of statistical data on selected exclusion threats (e.g. aging societies, number of people with disabilities, discrimination against women, etc.).

The selection of the above research methods is intended to provide a holistic approach to the issue of exclusion and constitute a comprehensive basis for developing principles for assessing inclusiveness and strategies for strengthening resistance to exclusion in Smart City.

The **originality** of the issue results from the following circumstances:

- focus on the issue of exclusion in Smart City, which is a less frequently discussed topic and falls within the less spectacular trend of analysing Smart City solutions;
- holistic analysis of exclusion taking into account various causes and types of this phenomenon;
- use of triangulation of research methods, including analysis of the role of exclusion in Smart City rankings, case studies of best practices in the field of inclusiveness and analysis of statistical data

for a targeted assessment of the scale and intensity of future threats related to exclusion;
- presentation of an original proposal for assessing inclusiveness and a strategy to strengthen resistance to exclusion in the Smart City.

It should also be added that the issue of exclusion in Smart City is important because the European Union promotes social cohesion as one of its main goals. Research of the phenomenon of social exclusion allows identifying areas of high risk and the need for intervention to correct social inequalities.

The monograph contains practical and theoretical content and concerns an important aspect of the quality of life in the city. Therefore, it boasts a wide group of potential readers. It may be particularly useful for researchers dealing with Smart City issues as well as teachers and students of fields related to city management and sociology. It can also be a teaching aid for urban planning and urban infrastructure.

Additionally, due to the methodology for assessing inclusiveness and the strategy of strengthening resistance to exclusion, the monograph may be interesting and useful for local and regional authorities. Finally, its recipient can be any urban stakeholder interested in the quality of life in the city.

1 The concept of a Smart City in modern economies

1.1 Creation and assumptions of the Smart City concept

The genesis of the Smart City concept goes deep to the roots of dynamic technological development and the increase in urbanization in the second half of the 20th century. Growing challenges related to the rapid growth of urban population, growing infrastructure needs and environmental pollution have prompted communities and decision-makers to look for innovative solutions. The Smart City concept has emerged as a response to these challenges, integrating modern information, telecommunications, energy and transport technologies to create more efficient, sustainable and citizen-friendly environments.

The Smart City concept is a response to global demographic changes and ongoing urbanization processes. It is the result of the development of an innovation ecosystem and the pursuit of sustainable progress. The first step in the development of a Smart City is to define challenges and priorities aimed at effectively solving these problems, which leads to the creation of the so-called Smart Cities. This term is usually associated with cities that have adopted a development strategy based on technology, creativity, openness to innovation and flexibility, understood as the ability to quickly adapt to changing external and internal conditions (Khavarian-Garmsir and Sharifi, 2022). In this context, various communication channels are used, such as e-governance or e-democracy (Makieła et al., 2022).

Arun was the scientist who first defined the Smart City concept in 1999. According to him, the foundation of this idea was information

DOI: 10.4324/9781003499992-2

technology, and an integral part of it was the commitment to improve the quality of life of an ordinary citizen (Arun, 1999). The creation of this definition coincided with the intensive development of the Smart City idea in the following years.

In the initial stage of development of the Smart City concept, the main focus was on the use of information technologies in managing urban infrastructure, improving access to public services and increasing energy efficiency. Monitoring systems, automation of urban processes, and the development of smart energy and transportation networks were key elements of this approach (Lyu and Hao, 2020).

Over time, the Smart City concept has evolved to take into account an increasingly wider range of elements, such as social participation, social innovation and even cultural aspects. The development of IoT and data analysis have become key elements, allowing monitoring and optimization of various aspects of urban life in real time.

Today's understanding of the Smart City concept also includes a strong emphasis on sustainable development, effective use of natural resources and the creation of resident-friendly places, both in terms of comfort and safety. The idea of Smart Cities is constantly being developed by technological progress, the growing needs of society and the desire to create better living conditions in cities, which makes it a dynamic area of research and implementation (Toli and Murtagh, 2020).

The history of Smart City development is a testimony to the evolution of society, technology and urbanization over the last few decades (Madhee, 2024). As already mentioned, the origins of this concept date back to the second half of the 20th century, when the dynamic growth of the urban population and the related challenges required innovative approaches to city management. In the 1960s and 1970s, cities began to move towards the use of computerization in administrative processes. The first IT systems were aimed at improving the efficiency of urban infrastructure management, such as transport and waste management. However, it was in the 1980s and 1990s that the real breakthrough occurred when information technology began to integrate with other fields, such as telecommunications, energy and transport (Vasudavan et al., 2019).

With the development of the Internet, the Smart City concept began to include the ideas of intelligent monitoring systems that enabled data collection in real time. This, in turn, allowed for a better understanding and optimization of the functioning of cities. In the 1990s, the first pilot projects were developed, where various solutions were tested in practice.

With the advent of the 21st century, the IoT has become a key element of the Smart City concept. It acted as the core, allowing communication and collaboration between various city devices. This combination of information, telecommunications and sensor technologies has enabled a more comprehensive approach to city management (Picioroaga et al., 2018).

The modern concept of a Smart City is not only technology, but also a complex ecosystem covering a wide range of fields (Ffitra et al., 2023). Developments in data analysis, AI and predictive algorithms are enabling more advanced forecasting and planning of urban development. In addition, the concept of a Smart City has also evolved to include sustainability, public participation and the creation of citizen-friendly places.

Given modern challenges, such as climate change, growing urban populations and the need for efficient resource management, the Smart City concept is becoming not only an innovative approach, but also an indispensable tool for creating better living conditions in cities. Its development is dynamic, driven by continuous progress (Mkrtychev et al., 2018).

Table 1.1 shows the key stages in the development of the Smart City concept and the characteristics of each of them. From the early days, when IT began to be introduced into urban management, through the integration of technology in the 1980s and 1990s, to the contemporary approach based on data analysis, AI and sustainability. The last 2 phases also underscore the growing importance of active public participation and the creation of places that respond to residents' needs.

The Smart City concept revolves around the 4T potentials: Tolerance, Trust, Technology and Talent. The city's progress in these areas dictates its level of intelligence, entrepreneurship and innovation, as highlighted by Makieła et al. (2022). Active engagement with the 4T factors plays a crucial role in the effective management of a Smart City, influencing the quality of life for residents and enhancing competitiveness within a larger entity, such as a metropolis. Beyond technology as a developmental support, there are other highly significant factors in promoting sustainable development. The description of all 4 factors is in Table 1.2.

The modern concept of a Smart City development is divided into 5 periods – from Smart City 1.0 to Smart City 5.0 (Figure 1.1). Each

Table 1.1 Characteristics of the early stages of Smart City development

Stage	*Characteristics*
Introduction of information technology in urban management	The use of computers in the 1960s and 1970s to improve the efficiency of administrative processes and infrastructure management
Integration of information technology in the 1980s and 1990s	Dynamic development of information technology, its integration into the telecommunications, energy and transportation sectors. Use of the first monitoring systems
Pilot projects in the 1990s	Beginning of practical experiments with various Smart City solutions in selected areas. Testing the capabilities and effectiveness of new technologies
The era of the IoT in the 21st century	Use of IoT in the development of a Smart City, enabling communication and cooperation between urban devices. Dynamic development of sensor systems
Advanced data analysis and AI	Introduction to the Smart City: advanced data analysis, use of AI and predictive algorithms for more efficient city management
Sustainability and public participation	Expanding the Smart City concept to include aspects of sustainability, active public participation and the creation of citizen-friendly places

Source: Authors' own study based on Sharifi and Alizadeh (2023); Khavarian-Garmsir and Sharifi (2022); Lyu and Hao (2020); Vasudavan et al. (2019); Picioroaga et al. (2018); Mkrtychev et al. (2018).

Table 1.2 Four T's of a Smart City

T type	*Characteristics*
Tolerance	The area of tolerance and diversity management is a captivating domain within management science. Tolerance involves embracing an uncritical understanding of individuals and appreciating their unique traits and features. These traits include age, origin, race and sexual orientation (Embarak, 2022). The evolution of diversity management initially focused on ensuring equal opportunities for ethnic and social minorities. Subsequent phases extended to equal

(*Continued*)

Table 1.2 (Continued)

T type	*Characteristics*
	treatment in employment and expanded to encompass interactions with customers, service recipients and various social groups. A city that fosters openness and tolerance stands a better chance of developing and achieving a higher level of social inclusion than one lacking these features.
Trust	The level of risk in cities and regions is on a rapid ascent, particularly in developing countries, where urban development may lack orderliness. Consequently, new technologies present an opportunity to enhance urban safety. Many cities have adopted ICT-based systems to bolster citizen security, where video surveillance is a primary system. Cities are shifting their urban innovation systems from traditional to innovative models like 'green', 'smart' and 'open', striving for environmental and social sustainability.
Technology	The integration of cutting-edge technologies, such as the IoT, AI, cloud computing, and data analytics empowers cities to efficiently manage resources, optimize infrastructure and provide innovative services. Smart Cities leverage technology to create intelligent networks that facilitate real-time communication and data exchange between various components, from smart traffic management systems to energy-efficient utilities and advanced public services. This interconnected ecosystem allows cities to make informed decisions, respond promptly to challenges and enhance sustainability.
Talent	Knowledge management forms the foundation of contemporary organizational management. Whether it involves examining market outcomes for globally or locally influential organizations, their strategies, developmental directions, market offerings and marketing approaches, they all underscore knowledge as a key determinant of modern thinking. Drawing from Kinelski's research on enterprises (Kinelski 2022), it is evident that the 'suppliers' of intellectual capital are creative residents, who are identified and managed through specialized programmes and methods rooted in the concept of human capital applied to the city as an organization. Based on their objectives and personnel programmes, these individuals are classified as talented artists, creative leaders, dedicated volunteers and more. Continuous, comprehensive development of innovation and creativity, along with the integration of key residents into the city, assumes paramount importance.

Source: Authors' own work based on Kinelski (2022).

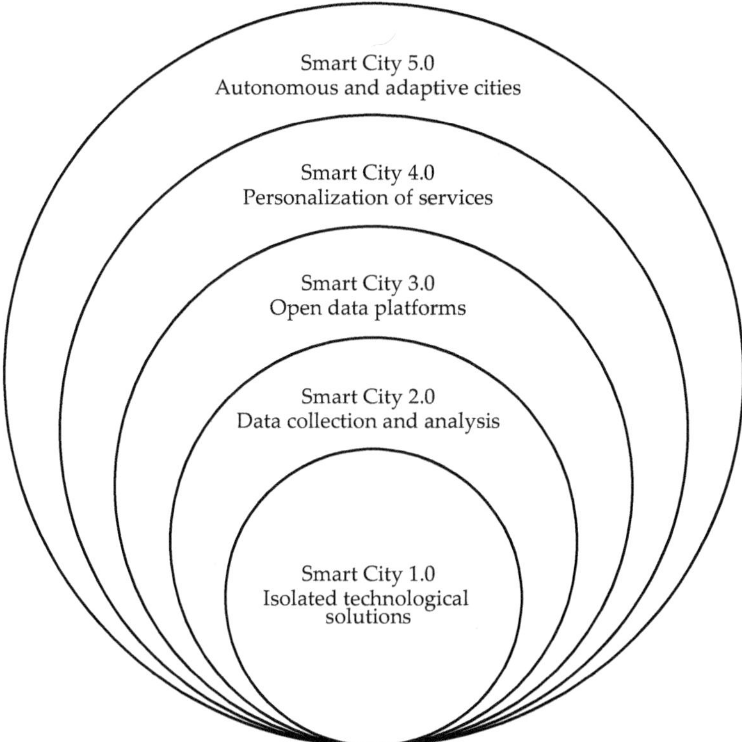

Figure 1.1 Five generations of Smart Cities.

Source: Authors' own study.

period brings the development of the Smart City concept with further important aspects. **Smart City 1.0** is the initial stage of Smart City development, where the implementation of modern technologies was initiated mainly by ICT companies. These companies introduce a variety of solutions, regardless of the actual needs of cities. An example of this approach is Songdo in South Korea, which is currently in the construction phase of a ubiquitous city. It is the largest private development project in the world, with the assumption that it will become a business centre comparable to such metropolises as Shanghai, Hong Kong, Kuala Lumpur and Singapore (Hussain et al., 2021). The initial period of Smart Cities was focused on individual, siloed solutions. Sensor-based traffic management, smart lighting and public WiFi were first to be adopted, introduced by companies keen to showcase their innovations. While efficiency gains were achieved, these isolated implementations lacked a centralized vision and often overlooked citizen needs.

Smart City 2.0 represents a stage in the development of Smart Cities, where public administration plays a key role (Trencher, 2019). In this case, it is the local government that initiates the use of modern technologies, and the introduction of innovative solutions is aimed at improving the quality of life of residents. In this period, data centres and centralized dashboards allowed for coordinated management of various systems, leading to improved traffic flow, waste management and energy consumption. However, concerns about data privacy and citizen involvement began to surface (Etezadzadeh, 2015). Examples of such cities include Seoul and San Francisco, which have been highlighted as examples of Smart City implementation (Lee et al., 2014). Today, most of the cities implementing Smart City projects are within Generation 2.0 (Azkuna, 2012).

Smart City 3.0 means that many influential modern cities are actively involving their residents in shaping further development. In this phase, local governments are focusing on creating spaces and opportunities to tap the diverse potential of residents as co-producers of public services. The focus shifted from technology-centric to citizen-centric (McKenna, 2016). Open data platforms empowered citizens to access information and engage in decision-making. Participatory budgeting and interactive platforms fostered dialog between residents and city authorities (Yun and Lee, 2020). However, challenges remained in ensuring equitable access to technology and addressing digital divides.

Although Smart City 3.0 remains focused on the use of modern technology to improve the quality of life in cities, its scope of interest also includes social, equality, educational and environmental issues, going beyond the projects characteristic of the second generation. Smart City 3.0 fits into the growing sharing economy (Nugraha, 2020). This often implies the need for courage on the part of city governments, which must confront the growing role of residents, for example by introducing participatory budgeting. Nevertheless, it is important not only to change the mentality (authorities-citizens), but especially the sphere of communication. At this stage, tools based on AI are beginning to be used more and more in the Smart City.

At the stage of **Smart City 4.0**, with the rise of AI and advanced analytics, cities are gaining access to deeper insights from their data. Predictive maintenance, AI-powered traffic management and personalized citizen services are becoming a reality. However, ethical considerations regarding AI usage and potential job displacement require careful navigation (Alexopulos et al., 2022). At this stage, customization and personalization of the city's services is being pursued.

In the development of a Smart City 4.0, it is essential to consider the intricate web of connections that yield tangible advantages. Employing blockchain technology for data encryption and distribution becomes crucial, as certain existing Smart City systems already leverage such technological advancements (Makieła et al., 2022). Building a Smart City 4.0 is closely tied to previous industrial revolutions characterized by robotics, AI, nanotechnology, the IoT and self-driving vehicles (Smart City 4.0, 2023). This transformative technological shift carries substantial social and economic implications for urban areas and the environment, integral to the pursuit of sustainable development, thereby establishing elevated expectations for citizens. Table 1.3 provides a description of differences between Smart City 3.0 and Smart City 4.0.

Table 1.3 Differences between Smart City 3.0 and Smart City 4.0

Feature	*Smart City 3.0*	*Smart City 4.0*
Focus	Citizen-centric	Data-driven and AI-powered
Key technologies	Open data platforms, participatory budgeting, digital twins	AI, predictive analytics, blockchains
Data usage	Used for transmitting informing decisions and personalized services	Used for deeper insights, prediction and automation
Citizen engagement	Active participation and feedback encouraged	Extensive analysis of needs and expectations
Partnership	Collaboration with some external partners	Extensive collaboration with various stakeholders
Challenges	Equitable access to technology, digital divide	Ethical considerations of AI, potential job displacement
Sustainability	Considered, but not the main focus	A core principle, integrated with solutions
Examples	Seoul, San Francisco	Cities investing heavily in AI and advanced analytics

Source: Alexopulos et al. (2022); Nugraha et al. (2020); Makieła et al. (2022), Garau et al. (2023).

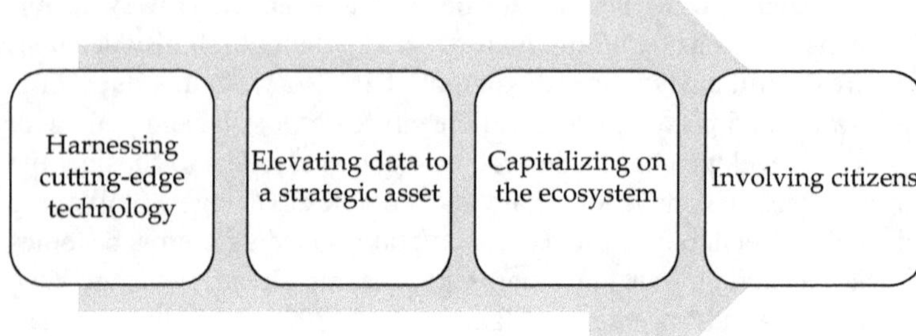

Figure 1.2 Steps towards Smart City 4.0.

Source: Authors' own study.

To reach Smart City 4.0, the following steps must be taken (Figure 1.2; Smart City 4.0, 2022):

- Harnessing cutting-edge technology. Smart City 4.0 has invested significantly in fundamental technologies, like cloud computing, mobile applications, IoT and Robotic Process Automation (RPA). Additionally, they have adopted specialized solutions, such as biometrics, blockchains, AI and telematics. These cities utilize smart technologies and applications to promote the achievement of Sustainable Development Goals (SDGs).
- Elevating data to a strategic asset. Smart City 4.0 leverages a diverse range of data types, with a particular emphasis on actively utilizing data from IoT, citizen engagement, biometrics, geospatial sources, peer-based interactions, business operations and real-time information.
- Capitalizing on the ecosystem. Smart City 4.0 collaborates extensively with external partners beyond local government entities. These partnerships include academic institutions, community organizations, industry associations, city networks, businesses and financial institutions.

- Involving citizens. More than half of Smart City 4.0 analyses the needs and expectations of citizens, actively seeking their feedback. Additionally, over 6 out of 10 cities closely collaborate with their employees to understand their requirements.

Smart City 5.0 represents an advanced stage in the evolution of cities, acknowledging their intricate and complex nature as urban systems (Svitek and Kozhevnikov, 2023). As cities continue to grow, they face challenges, such as traffic congestion, pollution, immigration, overcrowding and insufficient services. To address these issues, a new concept is proposed, viewing each city as a dynamic and continually evolving complex adaptive system (Rastogi et al., 2023).

Smart City 5.0 addresses the limitations of previous Smart City research, shifting from a provider-centric perspective (focusing on how technologies improve city objectives) to a citizen-centric viewpoint. It emphasizes how technologies can overcome constraints for citizens, such as the implementation of suburban working hubs to eliminate the need to navigate congested city centres (Becker et al., 2023). In essence, Smart City 5.0 aspires to create a 'liveable' city that goes beyond tangible benefits, connecting to the broader idea of a 'happy city' (Brdulak and Brdulak, 2017).

Smart City 5.0 represents a paradigm envisioning Smart Cities with a strong emphasis on human-centric design and addressing challenges faced by citizens (Svitek et al., 2020). This approach is akin to Industry 5.0, highlighting a human-centric philosophy that prioritizes core human needs and interests in city management (Svitek et al., 2023). Rather than solely exploring the possibilities of new technologies, Smart City 5.0 focuses on what the city can do for its residents, aiming to remove constraints that impact their perception of liveability.

Aligned with the European Commission's concept of Industry 5.0 (Breque et al., 2021), Smart City 5.0 aims to establish a living environment devoid of constraints by integrating human-centric principles into city solutions (Rastogi et al., 2022). The ultimate objective is to enhance well-being, as defined by the Organization for Economic Cooperation and Development (OECD), encompassing local material conditions, quality of life and sustainability (OECD, 2020). Smart City 5.0 strives to achieve well-being through initiatives that improve health, education, social services, citizen participation, environmental impact, vulnerability reduction and enhanced security (Becker et al., 2023).

Table 1.4 provides a characterization of the discussed Smart City 1.0–5.0 concepts, summarizing the current discussion.

At present, successive models of economic helixes clarify the concept and principles of cooperation between local participants in innovation processes in the face of the challenges of the global economy and the expectations of society. They are closely correlated with the development of smart urban structures described earlier. In retrospect, the following generations of helixes can be distinguished (Carayannis and Campbell, 2009; Carayannis et al., 2012):

- First Helix – Academia, universities, higher education system;
- Second Helix – Industry firms, economic system;
- Third Helix – State government, political system;
- Fourth Helix – Media-based and culture-based public;
- Fifth Helix – Natural environment, sustainability.

The innovation model, known as the Triple Helix, involves a series of interactions among the academia (represented by the university), industry and government. Its purpose is to promote economic and social development in alignment with concepts like the knowledge economy and knowledge society. In the theoretical framework of the innovation helix, each sector is depicted as a circle, or helix, with overlapping regions indicating interactions (Lopes and Luciano, 2023).

The initial model has evolved from 2 dimensions to illustrate more intricate interactions, particularly over time. The concept was originally formulated by Henry Etzkowitz and Loet Leydesdorff in the 1990s, with the release of 'The Triple Helix, University-Industry-Government Relations: A Laboratory for Knowledge-Based Economic Development' (Etzkowitz and Leydesdorff, 1995). The interactions between universities, industries and governments have led to the emergence of new intermediary institutions, such as technology transfer offices and science parks. Etzkowitz and Leydesdorff proposed a theoretical framework to explain the relationship between these 3 sectors and the formation of these innovative hybrid organizations. The adoption of the Triple Helix innovation framework has been widespread, and its application by policy-makers has played a role in transforming each sector.

Table 1.4 Evolution of Smart City from 1.0 to 5.0

Stage	*Time frame*	*Focus*	*Key features*	*Description*
1.0	1990s–2000s	Basic automation and efficiency	Traffic light control, automated waste collection	Isolated technology solutions; focus on improving operational efficiency; limited citizen engagement
2.0	2010s	Data collection and analysis	Sensors, IoT devices, integrated platforms, citizen engagement	Centralized platforms for data collection and management; improved resource allocation; early stages of citizen involvement
3.0	2020s	Human-centric	Personalized services, advanced technologies (blockchains, digital twins), ethical considerations, data privacy	Personalized citizen services; integration of advanced technologies like blockchains and digital twins; emphasis on ethical AI and data privacy
4.0	2025s	The usage of AI in a Smart City	Predictive analytics, sustainability, resilience, public–private partnerships	Open data platforms; AI-powered insights for decision-making; focus on sustainability and resilience; increased citizen participation
5.0	2030s+	Autonomous and adaptive cities	Seamless physical and digital integration, advanced AI and robotics, long-term sustainability, social equity	Self-healing infrastructure; autonomous mobility solutions; seamless integration of physical and digital worlds; focus on long-term sustainability and social equity

Source: Authors' own study.

The Triple Helix model states that the 'innovation ecosystems' within cities evolve through the collaboration of 3 distinct agents (Pique, 2019):

- Universities, acting as magnets to foster scientific and technological knowledge.
- The industrial sector, playing a crucial role in driving economic growth.
- Government entities at various levels (local, regional, national and international), actively involved in initiatives, governance and the formulation of land use policies.

The quadruple and quintuple innovation helix framework delineates interactions between universities, industries, government, public and environment within the context of a knowledge economy. Elias G. Carayannis and David F. J. Campbell (Carayannis and Campbell, 2009) co-developed the quadruple and quintuple innovation helix framework, with the Quadruple Helix model introduced in 2009 and the Quintuple Helix in 2010 (Carayannis and Campbell, 2010). The concept of extending the Triple Helix model to a Quadruple Helix was explored by various authors during the same period. The Carayannis and Campbell Quadruple Helix model integrates the public through the concept of a 'media-based democracy', highlighting the importance of effective communication of innovation policy by the government to the public and civil society through the media to garner support.

For industries engaged in research and development (R&D), the framework underscores the need for companies' public relations strategies to navigate media-driven 'reality construction'. The Quadruple and Quintuple Helix frameworks can be explained in terms of the knowledge models they utilize and the 5 subsystems (helixes) they encompass (Kuzior and Kuzior, 2020). In a Quintuple Helix model, knowledge and know-how are generated, transformed and circulated, influencing the natural environment (Suzic et al., 2020). Socio-ecological interactions facilitated by the Quadruple and Quintuple Helixes can be leveraged to identify opportunities for the knowledge society and knowledge economy, particularly in addressing sustainable development, including challenges like climate change.

The Quadruple Helix model emphasizes the inclusion of the cultural pillar and underscores the significance of constructing and communicating 'public awareness', asserting its impact across all dimensions of the system (Schütz et al., 2019). Nordberg (2015) characterizes the fourth helix as a more 'cultural' dimension and a backdrop crucial to the innovation roadmap. Ivanova (2014), approaching the subject from a systemic perspective with a focus on services, argues that the Quadruple Helix model not only engages consumers, but also addresses communication and media aspects (Yazici, 2023).

On a contrasting note, Höglund and Linton (2018) contend that the fourth helix is not a separate additional component, but an integral part of society, serving to respond to citizens' needs (Gasco-Hernandez et al., 2022). Despite the unquestionable contribution of the Quadruple Helix model, there exists a methodological challenge in how citizens introduce their public perspective, and how various actors define their functional role within society as a fourth pillar, collaborating with innovative processes (Taratori et al., 2021).

Carayiannis et al. (2012) describe the Quintuple Helix model as simultaneously interdisciplinary and transdisciplinary, highlighting the fifth helix's role in providing a more analytical perspective for the dynamic involvement of all stakeholders. This evolved model places a significant emphasis on an active and more 'human-oriented' approach, particularly emphasizing the circular process of knowledge production between subsystems, such as society and the economy (Crumpton et al., 2021). The Quintuple Helix model underscores the importance of collective efforts and exchange across education, the economy, the environment, society and political systems (Carayannis and Campbell, 2010).

The innovation processes within the Quintuple Helix model incorporate a fifth dimension, focusing on the nature and social ecology. However, a lingering question within this model is the challenge of connecting the 5 helixes in an innovation process. Markard et al. (2012) propose a response centred around 'ecology', which encompasses interdisciplinary relations between living organisms (social) or between them and their environments (natural), forming an integrated 'ecosystem'. The various concepts within the Quintuple Helix model converge in the society–nature transition (Van Den Berg and Verster, 2023). Consequently, the Quintuple Helix places particular emphasis on translating environmental and ecological issues, identifying them as crucial 'drivers' for future knowledge and innovation (Taratori et al., 2021).

The concept of successive Smart Cities from 3.0 to 5.0 is closely linked to successive helixes of economic development. The three-layered structure of economic development (Triple Helix) involves business, city and science and is gradually joined by the local community (Quadruple Helix) and environmental and cultural organizations (Quintuple Helix). This configuration of cooperation should ensure the achievement of the intended goals of all stakeholders and promote sustainable and dynamic economic development in the local and regional area.

Currently, 3 models of socio-economic cooperation – Triple Helix, Quadruple Helix and Quintuple Helix – are an important part of the evolution of the Smart City concept from version 3.0 to 5.0. In the context of Smart City 3.0, where AI and machine learning are beginning to play a key role, the Triple Helix model points to the integration of business, the city and science. Collaboration between these 3 sectors is becoming crucial for the effective use of modern technology to improve the quality of life of residents.

The transition to Smart City 4.0, which emphasizes human-centred design, personalization of services and ethical aspects, involves an expansion of the collaboration model. Quadruple Helix is including the local community, reflecting the growing role of citizens in decision-making. Together with business, the city and science, the local community is becoming an active participant in shaping dynamic and sustainable solutions for cities.

As we reach the Smart City 5.0 stage, where cities are becoming autonomous and adaptive, the Quintuple Helix model is becoming a key element. It includes not only business, city, science and community, but also environmental organizations. This expanded cooperation takes into account not only technological and social aspects, but also issues of sustainability and social equality. The efforts of these 5 areas of cooperation are aimed at creating autonomous and sustainable cities, integrating the physical and digital worlds and nurturing long-term social equality.

1.2 Areas of Smart City development and evaluation

The literature identifies 6 so-called pillars of Smart City (Table 1.5). Those pillars were first described by McKinsey Group in 2018 (McKinsey Global Institute, 2018). Smart governance plays a crucial role in leveraging digital technologies to streamline administrative

Table 1.5 Six pillars of Smart City

Pillar	*Description*	*Examples*
Smart Governance	Utilizes advanced information and communication technologies to streamline administrative processes, enhance transparency and improve decision-making within municipal authorities	Electronic voting systems, online citizen engagement platforms, digital services
Smart Economy	Fosters economic growth by integrating digital innovations, supporting entrepreneurship, encouraging innovation and implementing sustainable business practices	Innovation hubs, digital payment systems, technology incubators
Smart Mobility	Revolutionizes urban transportation through intelligent solutions, including efficient public transit, smart traffic management and promotion of sustainable transport modes	Intelligent traffic lights, real-time public transportation tracking, bike-sharing systems
Smart Environment	Monitors and manages environmental resources by optimizing waste management, conserving energy and ensuring sustainable water usage to minimize ecological impact	Smart waste bins, energy-efficient street lighting, water usage monitoring systems
Smart Living	Improves the quality of life for residents by deploying smart infrastructure, such as intelligent buildings, healthcare systems and educational institutions	Smart homes with automation, telemedicine services, smart classrooms
Smart People	Empowers citizens through digital literacy, access to information and community engagement, fostering a sense of inclusion and participation in shaping the city's future	Digital education initiatives, community engagement platforms, citizen forums

Source: Embarak (2021); Khan et al. (2020); Wolniak and Jonek-Kowalska (2023).

processes, enhance transparency and improve decision-making within municipal authorities (Embarak, 2021). This pillar focuses on using data-driven insights and communication technologies to create a more responsive and efficient government. The smart economy pillar is centred on fostering economic growth through the integration of digital innovations. This involves supporting entrepreneurship, encouraging innovation and implementing sustainable business practices (Izbash et al., 2023). By leveraging technology, Smart Cities aim to create a resilient and dynamic economic environment (Belaid et al., 2024).

Smart mobility is a key pillar that focuses on revolutionizing urban transportation. This involves the integration of intelligent transportation systems, efficient public transit options and the promotion of sustainable modes of transport (Wolniak, 2023). The goal is to reduce traffic congestion, enhance connectivity and minimize the environmental impact of transportation. In the case of the smart environment, cities strive to monitor and manage their environmental resources effectively. This includes deploying technologies to optimize waste management, conserve energy and ensure sustainable water usage. The emphasis is on creating urban spaces that minimize ecological impact and promote environmental sustainability (Xu, 2023).

Smart living is another important pillar that seeks to improve the quality of life for city residents. This involves the deployment of smart infrastructure, such as intelligent buildings, healthcare systems and educational institutions (De Matos and Ramos, 2023). These initiatives aim to create safer, more comfortable and accessible living environments for citizens. Lastly, the smart people pillar focuses on empowering citizens through digital literacy, access to information and community engagement (Khan et al., 2020). By fostering a sense of inclusion and participation, Smart Cities aim to create a more informed and actively involved citizenry, encouraging collaboration in shaping the city's future.

Implementing a Smart City is a complex process that is closely linked to the development of technical infrastructure. A key aspect is the development of advanced communications systems, including broadband Internet networks and cellular technologies. Telecommunications infrastructure plays a key role in enabling high-speed data transfer between different elements of a Smart City (Sharifi and Alizadeh, 2023).

Today's Smart Cities are complex ecosystems that use a range of advanced technologies to improve the quality of life of residents, increase the efficiency of urban services and contribute to sustainable development. One of the key elements of this concept is the development of the IoT, enabling various urban devices to connect and collaborate (Streitz et al., 2023).

Sensors and monitoring systems are an important component of the Smart City, enabling real-time data collection. With them, cities can monitor traffic, air quality, energy consumption or waste volumes, allowing them to react quickly to changes and optimize different areas of urban life. Data analysis and AI technologies also play a key role. Advanced algorithms allow processing vast amounts of information, identifying patterns and forecasting events. This enables more effective planning for urban development, optimized traffic management, or even faster response to emergencies (Tamas and Dora, 2023).

Smart transportation systems are another important area of technology in the Smart City. The development of public transportation, car-sharing systems, traffic management platforms or the development of autonomous vehicles are improving transportation accessibility, reducing traffic jams and emissions (Wolniak, 2023; Kowalska and Wolniak, 2023).

Energy is a key area, where modern technologies are being used. Smart Grids, or smart energy networks, allow efficient management of energy supply, increase the share of renewable sources and minimize losses (Gajdzik et al., 2024). Energy infrastructure must be upgraded to integrate new technologies, such as smart grids. These enable optimal management of energy consumption, monitoring of the grid and introduction of renewables.

Smart Cities prioritize involving and empowering citizens in the management of energy. They encourage active participation in energy conservation, promote informed decision-making regarding energy usage and foster contributions to a sustainable future. Innovative strategies in electric systems provide citizens with tools, platform and real-time data access to monitor and control their energy consumption, enabling them to actively play a role in creating a more efficient and sustainable electric system. In the context of Smart Cities, energy systems encompass an integrated network of infrastructure, technologies and policies designed to facilitate efficient energy generation, distribution and consumption within urban environments. These systems aim to optimize energy usage, mitigate carbon emissions and enhance overall sustainability.

Adoption of a new approach to electric systems in Smart Cities becomes imperative to meet the increasing energy demands, address environmental concerns, bolster grid resilience, accommodate decentralized energy generation, leverage technological advancements and empower citizens. Through the integration of renewable energy sources, smart grids, energy storage solutions, as well as advanced monitoring and management systems, Smart Cities can establish electric systems that are more efficient, sustainable and resilient, tailored to the evolving needs of urban environments (Gajdzik et al., 2024).

Seven primary elements constitute the energy systems in Smart Cities: renewable energy generation, smart grids, energy storage, demand response and energy efficiency, electric mobility, energy management systems and citizen engagement. Table 1.6 provides detailed

Table 1.6 Main components of energy systems used in Smart Cities

Component	*Characteristics*
Renewable energy generation	Smart Cities prioritize utilizing renewable energy sources, such as solar, wind, geothermal and biomass. They integrate various renewable systems like solar panels, wind turbines and geothermal heating/cooling.
Smart grids	Advanced electrical grids in Smart Cities incorporate digital technologies, sensors and automation to optimize electricity distribution. Smart grids enable real-time monitoring, efficient load balancing and integration of decentralized energy sources. They also facilitate two-way communication between utilities and consumers, supporting demand response programmes and energy conservation.
Energy storage	Crucial for Smart Cities, energy storage technologies efficiently capture and store surplus energy from renewables during peak production. Solutions like batteries, pumped hydro storage and compressed air energy storage balance electricity supply and demand, ensure grid stability and provide backup power during outages.
Demand response and energy efficiency	Smart Cities implement demand response programmes to encourage consumers to adjust energy usage during peak demand. Real-time data and communication networks provide incentives for reducing or shifting energy consumption. Additionally, energy-efficient technologies and practices, including smart appliances, LED lighting, intelligent HVAC systems and building automation, are promoted.

Table 1.6 (Continued)

Component	*Characteristics*
Electric mobility	Smart Cities promote electric vehicle (EV) adoption through comprehensive charging infrastructure, strategically placed in residential areas, parking lots and public spaces. Integration with renewable sources and smart grids manages increased energy demand, promoting sustainable transportation options.
Energy management systems	These systems use data analytics and AI algorithms to monitor and optimize energy consumption across various sectors. They collect and analyse data from sensors and devices to identify inefficiencies, recommend energy-saving measures and enable predictive maintenance. Energy management systems allow real-time monitoring, analysis and control of energy usage at the city-wide level.
Citizen engagement	Smart Cities actively engage citizens in energy conservation. They provide real-time energy consumption data, user-friendly interfaces and personalized recommendations. Awareness campaigns, educational programmes and interactive platforms empower individuals and communities to make informed choices for a sustainable energy future.

Source: Gajdzik et al., (2024); Sarjana (2023); Darmawan et al. (2023).

characteristics of each of these components. The goal of integrating these elements and encouraging collaboration among stakeholders is to optimize energy utilization, decrease carbon emissions, improve grid resilience and cultivate a sustainable and enjoyable urban environment.

Smart City communication relies on telecommunication technologies, creating smart networks that enable the rapid exchange of information between devices, as well as between residents and city authorities. It is also worth highlighting the development of blockchain technologies, which can find application in the area of data security, transparency of transactions or even in urban identification systems (Kalenyuk et al., 2023). The implementation of Smart City also requires the modernization of building infrastructure to enable the integration of smart solutions into existing urban infrastructure.

Notably, the most important technologies in a Smart City are not just single solutions, but a comprehensive system, in which diverse technologies work together to create smart, integrated urban environments. It is the interaction of various innovations – from sensors to AI – that contribute to efficient and sustainable urban development. Table 1.7 provides an overview of selected technologies used in Smart Cities.

Table 1.7 Characteristics of selected technologies used in Smart Cities

Technology	*Characteristics*
IoT	The IoT enables the integration of various devices, sensors and systems to form a smart network. This allows data on the environment, transportation, energy and other areas to be collected and then analysed for effective city management. IoT can include smart street lights, environmental sensors, smart energy meters, etc.
Sensors and monitoring	Using advanced sensors to continuously monitor various parameters of the urban environment. Sensors can measure pollution levels, humidity, temperature, noise levels, as well as monitor traffic and waste volumes. This data is then analysed to make informed urban planning decisions.
Big Data Analytics	Big Data Analytics is a technology that makes it possible to analyse huge amounts of data, from various sources, to identify patterns and trends. In the context of the Smart City, data analytics allows for a better understanding of residents' behaviour, urban infrastructure needs and optimization of public service delivery.
Traffic management systems	Advanced traffic management systems include smart traffic signals, traffic monitoring cameras, dynamic traffic guidance systems and parking monitoring. These technologies help minimize traffic jams, improve road safety and optimize the use of urban space.
E-mobility	E-mobility focuses on the introduction of electric modes of transportation, such as electric cars, electric bicycles and scooters. Infrastructure supporting e-mobility includes charging stations and traffic management systems for electric vehicles. The goal is to reduce greenhouse gases emissions and improve air quality.
Smart grid	Smart Grid integrates communications technology into the power grid. This allows monitoring, control and optimization of electricity distribution in real time. Smart Grid helps increase energy efficiency and integrate renewable energy sources.

Table 1.7 (Continued)

Technology	*Characteristics*
Waste management systems	Waste management systems use technology to monitor the capacity and condition of trash containers, optimize trash collection routes and promote waste segregation. Smart garbage cans can report when they are full, enabling efficient waste management and minimizing environmental impact.
Digital participatory platforms	Participatory technologies include online platforms, mobile apps and social media tools that enable residents to actively participate in decision-making processes. This can include public consultations, online voting, participation in community projects and tracking government activities.
Smart lighting	Smart lighting systems use motion, time and weather sensors to adjust light intensity. This saves energy by turning the light on only when needed, contributing to a sustainable city.
Blockchains in government	Blockchain technology used in government can secure data, facilitate transactions, eliminate bureaucracy and increase transparency. With its decentralized structure, blockchains can improve data security and facilitate administrative processes, such as citizen data management and property registration.
AI	AI in Smart Cities can be used to analyse data, forecast trends, optimize traffic, personalize public services and improve energy efficiency. Machine learning algorithms can help respond more quickly to changes and adapt to residents' needs.
5G and mobile networks	The deployment of 5G technology and the development of high-speed cellular networks enable fast and stable data transmission, which is crucial for the efficient operation of many Smart City applications. This supports communication between IoT devices, enabling seamless city management and improving access to online services.
Biometrics and identification	Biometric technologies, such as facial recognition, fingerprint readers and iris scanners, can be used to secure access to buildings, monitor public safety and authorize transactions, helping to improve overall security in the city.
Drones and unmanned aerial vehicles	The use of drones in the Smart City includes monitoring traffic, patrolling urban areas, delivering packages and inspecting infrastructure. This acts as an additional tool for rapid response to emergencies, monitoring hard-to-reach areas and enhancing public safety.

(*Continued*)

Table 1.7 (Continued)

Technology	*Characteristics*
Augmented reality	Augmented reality (AR) technology can be used to provide information about the surrounding environment, urban navigation, education and to create interactive cultural experiences. AR can improve residents' interaction with the urban environment and enhance education and tourism processes.
Green technologies	They include innovative solutions for sustainable urban development, such as energy-efficient buildings, green roofs, photovoltaic panels and rainwater recycling. These technologies contribute to reducing emissions, protecting the environment and improving the quality of life for residents.
Epidemic tracking systems	Using technologies, such as GPS tracking, AI and data analytics, to monitor, predict and manage potential epidemics, which is crucial in the context of public health.
Smart health	Integration of technology in the health sector includes telemedicine, population health monitoring, medical data management systems and access to electronic patient records. Smart health supports efficient healthcare delivery, disease prevention and improved quality of life for residents.
Geolocation and GIS	Geographic information systems (GIS) and geolocation technologies help in effective urban planning, management of urban infrastructure data, analysis of areas at risk and creation of interactive maps for residents. These are tools that support sustainable development and urban planning.
Urban robotics	Urban robotics includes the use of robots for a variety of tasks, such as maintenance, deliveries, infrastructure inspections and construction work. Robots can operate autonomously or be remotely controlled, helping to improve efficiency and safety in the city.

Source: Authors' own work based on Wolniak (2023); Streitz et al. (2023); Kalenyuk et al. (2023); Sarjana (2023); Tamas and Dora (2023); Duan et al. (2022); Zhou (20220; Al Nuaimi et al. (2015); Wolniak and Grebski (2023); Kvalvik et al. (2023); Vlahkis et al. (2023); Dadwal et al. (2023); Barrera-Camara et al. (2023); Ruiz-Vanoye (2023); Pellatt and Palfreman (2023); Kondratenko et al. (2023); Caputo et al. (2015); Raj and Shetty (2024); Ali et al. (2023); Ddeja et al. (2023); Kramarz et al. (2022); Jonek-Kowalska (2022); Jonek-Kowalska (2023); Wolniak (2023); Stecuła et al. (2023).

Smart City and Industry 4.0 are interdependent areas that support each other, leading to more integrated and efficient societies. Industry 4.0, also known as the fourth industrial revolution, focuses on the intelligent use of digital technologies in manufacturing, while Smart City focuses on using these technologies to improve the quality of life in urban areas.

In the context of Industry 4.0, automation, cyber-physical systems, the IoT and data analytics are key. These same technologies are also key to Smart City. The automation of manufacturing processes in Industry 4.0 aims to increase efficiency, reduce costs and optimize resources. In Smart City, these technologies are used to monitor and manage various aspects of urban life, such as transportation, energy, waste management and public services (Kurshudov, 2024).

Industry 4.0 is bringing innovation to manufacturing, introducing smart factories, where machines communicate with each other and make autonomous decisions. Within the Smart City, the same principles can be applied to urban infrastructure management. For example, smart traffic management systems can use sensor data and data analysis algorithms to optimize traffic, reduce congestion and improve safety (Wolniak, 2023).

The IoT plays a key role in both Industry 4.0 and Smart City. In the context of Industry 4.0, devices in manufacturing plants communicate with each other, enabling monitoring, analysis and optimization of processes (Manfreda and Mijač, 2024). In Smart City, IoT enables the connection of various elements of urban infrastructure, such as smart streetlights, environmental sensors and Smart City platforms, to collect data, monitor and manage resources efficiently.

The collaboration between Industry 4.0 and Smart City is helping to create more sustainable, efficient and resident-friendly cities. Smart technologies not only streamline production processes, but also enable dynamic and adaptive management of urban infrastructure, resulting in improved quality of life for residents and optimal use of urban resources. As a result, this synergy between Industry 4.0 and Smart City is a key development direction for modern societies.

Table 1.8 presents a comparison between Smart City and Industry 4.0.

Implementation of the Smart City concept has the potential to affect public policy in the city. Using advanced digital infrastructure and data-driven solutions, the implementation of Smart City solutions can help improve security (Teng et al., 2024). One key aspect in

Table 1.8 Comparison between the Smart City concept and Industry 4.0

Aspect	*Smart City*	*Industry 4.0*
Scope	Encompasses a wide range of urban services and infrastructure, including transportation, energy, healthcare and governance.	Primarily focuses on the transformation of manufacturing and industrial processes, optimizing production through digital technologies.
Key technologies	IoT, Big Data, AI, sensors, smart grids and communication technologies are integral for managing city services efficiently.	IoT, Industrial IoT (IIoT), cloud computing, machine learning and robotics play crucial roles in creating smart and interconnected industrial systems.
Objectives	Enhancing quality of life, sustainability and efficiency in urban areas. Improving public services, infrastructure and citizen well-being.	Enhancing manufacturing processes, increasing efficiency, reducing downtime and enabling more flexible and agile production. Improving productivity and product quality.
Focus on data	Relies heavily on data analytics for optimizing city operations, predicting trends and improving decision-making in areas like traffic management and waste disposal.	Emphasizes data-driven decision-making in industrial settings to optimize production, predict maintenance needs and enhance overall operational efficiency.
Integration of systems	Aims to integrate various urban systems, such as transportation, energy and public services, to create a holistic and interconnected urban environment.	Focuses on integrating digital technologies into the entire manufacturing process, from design and production to supply chain and customer service.
User involvement	Involves citizen participation through platforms for feedback, engagement and collaboration to address the needs and preferences of residents.	Involves collaboration between machines, systems and human operators, emphasizing human-machine interaction and cooperation in the manufacturing environment.

Table 1.8 (Continued)

Aspect	*Smart City*	*Industry 4.0*
Cyber-physical systems	Emphasizes the deployment of sensors and devices in the physical environment to collect and transmit data for improved decision-making and automation.	Utilizes cyber-physical systems in manufacturing equipment and processes to enable real-time monitoring, control and optimization of industrial operations.
Challenges	Privacy concerns, security issues and the need for extensive infrastructure investment are common challenges.	Implementation costs, workforce adaptation and potential job displacement due to automation are challenges faced in the adoption of Industry 4.0.
Global adoption	Smart City initiatives are being implemented globally on various scales, with cities around the world investing in technology to become smarter and more sustainable.	Industry 4.0 is a global phenomenon transforming manufacturing industries, with countries and industries adapting to stay competitive in the digital age.

Source: Authors' own work based on Wolniak et al. (2023).

this regard is improving public safety through advanced monitoring systems. Smart City initiatives often include the deployment of intelligent video analytics, sensors and IoT devices to detect and respond to incidents in real time. This approach can help prevent crime, optimize emergency response and generally maintain public order (Li et al., 2024).

Traffic management is another important area, where Smart City implementation can have a significant impact. Smart transportation systems use data analysis to optimize traffic flow, reduce traffic jams and improve overall mobility. This not only leads to more orderly and efficient transportation, but also reduces the potential for traffic-related disruptions and conflicts (Elhofy et al., 2023).

When it comes to public management, Smart City systems use data analysis and digital platforms to streamline administrative processes. This can result in improved delivery of public services, faster responses to citizens' concerns and increased overall communication between authorities and residents.

The implementation of smart infrastructure in the city contributes to sustainable development, which indirectly affects public policy. Initiatives such as efficient waste management, energy conservation and pollution monitoring can contribute to a cleaner and healthier urban space. A clean and healthy environment can positively affect the well-being of residents and contribute to the quality of urban life (Rele and Patil, 2023).

In terms of public policy in Smart Cities, it is important to address potential challenges related to privacy, data security and equal access to these technologies to ensure that the benefits of Smart City implementation are felt by all residents. Furthermore, public awareness and involvement are key to the successful adoption of these technologies, as informed and engaged citizens are more likely to support and actively participate in Smart City development. Table 1.9 reviews selected Smart City solutions that impact security.

Table 1.9 Solutions used in Smart Cities that affect security

Area	*Description*
City monitoring	The use of advanced surveillance systems, smart cameras and sensors to continuously monitor the urban area. This allows for rapid detection of incidents, including crimes, accidents or dangerous situations.
Smart video analysis	The use of video analysis technology to identify unusual behaviour, facial recognition or vehicle registration. This allows rapid response to potential threats and tracking of suspects.
Alarm systems	Implementation of smart alarm systems that can be activated automatically in response to identified threats. This allows for instant notification of emergency services and minimizes response time.
IoT sensor networks	Expand the infrastructure of IoT sensors to monitor various parameters, such as air pollution levels, noise or water quality. This data is analysed in real time to maintain a safe environment.
Smart street lighting	The use of street lighting systems that adapt to environmental conditions. Smart streetlights can respond to movement by changing the intensity of the light, increasing safety and saving energy.

Table 1.9 (Continued)

Area	*Description*
Emergency management systems	Creating integrated emergency management platforms that result in a coordinated response to emergencies. This includes monitoring, communicating and coordinating emergency services and informing communities of risks.
Safe transportation infrastructure	Implementation of traffic monitoring systems that identify and manage road hazards. Smart traffic signals, cameras and warning systems help ensure safe mobility in the urban area.
Public safety mobile apps	Developing mobile apps to quickly report incidents, access safety information and receive alerts. This increases community involvement in maintaining safety through active participation in the monitoring and reporting process.
Analysing data to predict threats	Using advanced data analysis technologies to predict potential threats. Algorithms can analyse data from a variety of sources, identifying patterns and correlated events, making it possible to prevent incidents before they occur.
Education and public awareness	Organizing educational campaigns and activities to raise community awareness of the benefits and security associated with Smart City technology. Informing residents about measures taken to increase safety and promoting active participation.

Source: Li et al. (2024); Konstantopoulou et al. (2024); Erces et al. (2023); Osipova and Hornecker (2023); Elhofy et al. (2023); Qian (2023); Role and Patil (2023); Tomitsch et al. (2023); Vătăşoiu et al. (2023).

Smart City technologies have a significant impact on public participation, enabling more active involvement of residents in public life. Thanks to mobile applications, online platforms and other innovative solutions, the community has easier access to information and the opportunity for direct dialog with local authorities (Arora et al., 2024). Information technologies enable rapid communication of information about local issues, events or planned projects. Residents can raise issues, provide feedback and actively participate in the decision-making process through public consultations or online voting. Participatory platforms integrate different social groups, allowing them to exchange ideas and work together on local initiatives. With this, the community becomes more organized and able to effectively express its needs and expectations.

In addition, smart transportation or waste management solutions allow residents to actively participate in caring for the environment and make more informed decisions about the use of urban resources. As a result, Smart City technologies support the development of civil society, foster transparency in the actions of local authorities and allow for more democratic participation of residents in shaping their environment (Domingo et al., 2021). This makes public participation more dynamic, effective and accessible to a wider spectrum of the local community.

As part of the Smart City, mobile applications are being used in a variety of areas, aiming to improve the city's daily operations. An important application of these is to facilitate public transportation through apps that allow for route planning, online ticketing, monitoring delays or informing about schedule changes (Maddel et al., 2023). Therefore, residents can move around the city more efficiently and the traffic management becomes smoother. Mobile applications are also being used to monitor and manage urban infrastructure, such as street lighting, water supply systems and waste management (Neofotistos et al., 2023). Smart sensors and remote control allow efficient management of resources, minimizing energy consumption and increasing environmental performance.

In the area of public safety, mobile apps support city monitoring systems, allowing residents to report incidents and receive warnings or emergency information (Alam et al., 2022). This increases cooperation between the community and security authorities. In addition, mobile apps are being used to improve access to public services, such as healthcare, education and administration. Electronic appointment scheduling, remote medical consultations or educational platforms are just a few examples of how mobile apps support the operation of these sectors (Stahl and Reiterer, 2022).

From a Smart City perspective, mobile apps are an integrative tool that connects different aspects of urban life, fostering efficiency, sustainability and better alignment of the city with the needs of its residents. Table 1.10 reviews examples of mobile apps that influence social participation in Smart City.

An important issue related to Smart City development is its environmental impact. Smart Cities often promote sustainable transportation, favouring public transportation, electric bicycles or electric cars.

Table 1.10 Example of applications used in the Smart City context

Application	*Characteristics*
mObywatel	This national app allows users to access various public services electronically, including reporting problems, registering vehicles, purchasing public transport tickets and voting in local elections. It fosters participation by simplifying interactions with government and empowering citizens to engage in civic activities.
Locate Me	This app empowers people with disabilities by facilitating navigation and access to public services and facilities. It allows users to report accessibility issues and suggest improvements, promoting inclusive infrastructure and social participation.
DecydujMy	This platform enables citizens to participate in decision-making processes by engaging in public consultations, proposing ideas and voting on initiatives. It encourages open dialog and collaboration between residents and local authorities.
FIXiT	This app allows users to report problems in their city, ranging from potholes to vandalism. It enables citizens to contribute to improving their environment and fosters a sense of community responsibility.
Kulturalna Warszawa	This app provides information about cultural events in Warsaw, allowing users to discover activities, purchase tickets and share their experiences. It promotes cultural engagement and encourages social interaction.
Nextbike	This bike-sharing app facilitates sustainable transportation and allows users to find available bikes for rent. It promotes healthy lifestyles and active participation in urban life.
Lokal	This app provides real-time information about public transport schedules and delays. It improves mobility and accessibility, enabling citizens to plan their journeys and participate more actively in city life.
EcoMap	This app displays air quality data and encourages users to report and discuss environmental issues. It promotes environmental awareness and empowers citizens to advocate for sustainable practices.

Source: Authors' own work based on mObywatel (2024); Locate Me (2024); DecydujMy (2024); FIXiT (2024); Kulturalna Warszawa (2024); Nextbike (2024); Lokal (2024); SoS Alarm (2024).

This approach reduces emissions, improving air quality and reducing atmospheric pollution, which has a positive impact on residents' health and the environment.

In addition, Smart City waste management systems enable more efficient trash management. Monitoring waste levels and segregation automatically lead to a reduction in the amount of waste going to landfills, which reduces the negative impact on soil and water. The previously discussed smart technologies also support sustainable urban design, promoting urban greening, parks and the general availability of recreational areas. This contributes to improving the quality of life for residents and preserving biodiversity in the urban area.

However, it is worth noting that the introduction of the Smart City requires strict control and monitoring to avoid potential negative effects, which will be discussed in detail in Chapter 2, such as excessive consumption of resources for the production of modern technology or the generation of additional electronic waste. Ultimately, the development of Smart City can contribute to reducing the negative impact of cities on the environment, provided the systems are properly designed, implemented and maintained.

The economic development of a Smart City depends on a number of factors that combine to create innovative and sustainable communities (Jonek-Kowalska and Wolniak, 2021). The first key element, discussed in detail earlier, is the technological infrastructure, including broadband Internet connections, sensor networks and data management systems. This allows for the efficient use of information, which is key to improving various aspects of city life.

Another important factor is community involvement and cooperation between the public sector, private sector and civil society. The Smart City concept involves the active participation of residents in decision-making processes and the use of their knowledge in creating innovative solutions. Education and skills development are also a key element in supporting Smart City economic development. Investments in digital education and training related to modern technologies enable residents to use new tools and foster the creation of new jobs.

Sustainability is another aspect that affects the economic progress of a Smart City. This happens through efficient management of resources, reduction of CO_2 emissions, development of public transportation based on green technologies and promotion of renewable energy.

Private and public investments in the new technology sector and start-ups that create innovative solutions have a significant impact on the economic development of the Smart City (Jonek-Kowalska and Wolniak, 2021). Creating a favourable business environment that supports the development of new technologies attracts entrepreneurs and investors.

The economic importance in Smart City development can be observed following the research by Jonek-Kowalska and Wolniak (2021). They observed that the progress of Smart City initiatives in Poland faces challenges due to the inadequate economic well-being of residents and the challenging financial circumstances of cities, limiting their capacity to undertake Smart City projects. Polish cities show a preference for the Triple Helix model, concentrating on fostering business collaborations and establishing favourable conditions for commercial entrepreneurship, rather than engaging higher-order helixes that involve communities or ecological organizations. This inclination might contribute to urban challenges, such as unsustainability and socio-economic exclusion. The evaluated Polish cities are considered average in terms of financial stability, wealth, investment, administrative efficiency and citizen participation. The weakest aspect pertains to the wealth and quality of life of residents, including gross remuneration levels. The economic stability of the city and its indebtedness strongly impact investment parameters, subsequently influencing the city's ability to advance Smart City initiatives.

The underdevelopment of Polish Smart Cities is also evidenced by the predominant utilization of conventional financing methods, relying primarily on budgetary funds and sponsorships, rather than embracing more innovative approaches like public–private partnerships or participatory budgets. To enhance the progression of Smart Cities in Poland, it is imperative to provide support to less developed cities, alleviate local budgets from non-stimulative expenditures, address the fundamental needs of urban communities, improve demographic trends and establish transparent and accessible standards for private–public partnerships.

Bibliography

Al Nuaimi, E., Al Neyadi, H., Mohamed, N., Al-Jaroodi, J. (2015). Applications of big data to smart cities. *Journal of Internet Services and Applications*, 6(1), 1–15.

Alam, T., Hadi, A. A., Najam, R. Q. S. (2022). Designing and implementing the people tracking system in the crowded environment using mobile application for smart cities. *International Journal of System Assurance Engineering and Management*, 13(1), 11–33.

Alexopoulos, C., Keramidis, P., Pereira, G. V., Charalabidis, Y. (2022). Towards Smart Cities 4.0: Digital participation in smart cities solutions and the use of disruptive technologies. *Lecture Notes in Business Information Processing*, 437, 258–273.

Ali, S. A., Elsaid, S. A., Ateya, A. A., ElAffendi, M., El-Latif, A. A. A. (2023). Enabling technologies for next-generation smart cities: A comprehensive review and research directions. *Future Internet*, 15(12), 398.

Arora, A., Jain, A., Yadav, D., Chamola, V., Sikdar, B. (2024). Next generation of multi-agent driven smart city applications and research paradigms. *IEEE Open Journal of the Communications Society*, 4, 2104–2121.

Arun, M. (1999). Smart cities: The Singapore case. *Cities*, 16(1), 16–22.

Azkuna, I. (Ed.) (2012). Smart cities study: International study on the situation of ICT, *innovation and knowledge in cities*. Bilbao: The Committee of Digital and Knowledge-Based Cities of UCLG.

Barrera-Cámara, R. A., Fuentes-Penna, A., Bernabe-Loranca, M. B. (2023). Tools and technologies for smart education in sustainable smart cities. *Management, Technology, and Economic Growth in Smart and Sustainable Cities*, 156–173.

Becker, J., Chasin, F., Rosemann, M. (2023). City 5.0: Citizen involvement in the design of future cities. *Electron Markets*, 33, 10. https://doi.org/10.1007/s12525-023-00621-y

Belaid, F., Amine, R., Massie, C. (2024). Smart cities initiatives and perspectives in the MENA region and Saudi Arabia. In *Smart cities*. Studies in Energy, Resource and Environmental Economics, Berlin: Springer, 295–313.

Brdulak, A., Brdulak, H. (2017). *Happy city: How to plan and create the best livable area for the people.* Berlin: Springer. https://doi.org/10.1007/978-3-319-49899-7.

Breque, M., De Nul, L., Petridis, A. (2021). *Industry 5.0: Towards a sustainable, human-centric and resilient European industry.* Luxembourg: European Commission, Directorate-General for Research and Innovation. https://ec.europa.eu/info/publications/ (accessed 16.02.2024).

Caputo, F., Magliocca, P., Canestrino, R., Rescigno, E. (2015). Rethinking the role of technology for citizens' engagement and sustainable development in smart cities. *Sustainability*, 15(13), 10400.

Carayannis, E. G., Barth, T. D., Campbell, D. F. J. (2012). The Quintuple Helix innovation model: Global warming as a challenge and driver for innovation. *Journal of Innovation and Entrepreneurship*, 1, 2.

Carayannis, E. G., Campbell, D. F. J. (2009). 'Mode 3' and 'Quadruple Helix': Toward a 21st century fractal innovation ecosystem. *International Journal of Technology Management*, 46(3/4), 201–234.

Carayannis, E. G., Campbell, D. F. J. (2010). Triple Helix, Quadruple Helix and Quintuple Helix and how do knowledge, innovation and the environment relate to each other: A proposed framework for a trans-disciplinary analysis of sustainable development and social ecology. *International Journal of Social Ecology and Sustainable Development*, 1(1), 41–69.

Crumpton, C. D., Wongthanavasu, S., Kamnuansilpa, P., Draper, J., Bialobrzeski, E. (2021). Assessing the ASEAN Smart Cities Network (ASCN) via the Quintuple Helix Innovation Framework, with special regard to smart city discourse, civil participation, and environmental performance. *International Journal of Urban Sustainable Development*, 13(1), 97–116.

Dadwal, S. S., Jahankhani, H., Bowen, G., Nawaz, I. Y. (2023). *Technology and talent strategies for sustainable smart cities: Digital futures*. London: Emerald.

Darmawan, A. K., Muhsi, M., Anekawati, A., Umam, B. A., Jalil, D. K. A. (2023). An interpretive structural model approach to strategic management modelling for sustainable smart village development in Indonesia. In *10th International Conference on ICT for Smart Society*, 6–7 Sept. 2023, Bandung, Indonesia.

DecydujMy Razem – Gliwicka Platforma Partycypacyjna. (2024). https://decydujmyrazem.gliwice.pl/ (accessed 16.02.2024).

Deja, A., Ślączka, W., Dzhuguryan, L., Dzhuguryan, T., Ulewicz, R. (2023). Green technologies in smart city multifloor manufacturing clusters: A framework for additive manufacturing management. *Production Engineering Archives*, 29(4), 428–443.

deMatos, N., Ramos, C. M. Q. (2023). Smart cities: Operational concepts for an elusive framework. In *Handbook of research on network-enabled IoT applications for smart city services*, Hershey, PA: IGI Global, 323–338.

Domingo, J., Cabello, K. A., Rufino, G. A., Villanueva-Jerez, M. J., Sarmiento, C. J. (2021). A framework in developing a citizen-centered smart city mobile application as a platform for digital participation in iloilo city. *International Archives of the Photogrammetry, Remote Sensing and Spatial Information Sciences*, 46(4/W6-2021), 153–160.

Duan, Y., Xu, C., Hu, W., Zhao, W. (2022). Research on the regulation of artificial intelligence on the orderly development of China's intelligent economy. In *2022 Asia-Pacific Computer Technologies Conference*, Wuhan, China, January 7–9, 2022, 11–15.

Elhofy, M., Abdelaziz, M., Omran, I., Abdelwahab, M. (2023). Effects of independent wheels steering system on vehicle cornering performance and road safety of the smart cities. *Ain Shams Engineering Journal*, 14(6), 102097.

Embarak, O. (2021). Smart city transition pillars with layered applications architecture. *Procedia Computer Science*, 191, 57–64.

Embarak, O. (2022). Smart cities new paradigm applications and challenges. In *EAI/Springer Innovations in communication and computing*, Berlin: Springer, 147–177.

Érces, G., Rácz, S., Vass, G., Varga, F. (2023). Fire safety in smart cities in Hungary with regard to urban planning. *Journal of Integrated Disaster Risk Management*, 13(2), 104–128.

Etezadzadeh, C. (2015). *Smart city – Future city? Smart City 2.0 as a livable city and future market.* Berlin: Springer.

Etzkowitz, H., Leydesdorff, L. (1995). The Triple Helix: University-industry-government relations – A laboratory for knowledge based economic development. *Glycoconjugate Journal*, 14(1), 14–19.

Fitria, I., Rosa, H. U., Firmansyah, M., Lembang, G. R., Mardianto, M. F. F. (2023). Analysis of factors influencing public perceptions of smart city concept of Indonesia's new capital city. *AIP Conference Proceedings*, 2975(1), 080016.

FIXiT. (2024). https://play.google.com/store/apps/details?id=nz.co.android.touchtech&hl=pl (accessed 16.02.2024).

Gajdzik, B., Wolniak, R., Grebski W., Grebski, M., Danel, R. (2024). *Smart cities with smart energy systems. Key development directions.* Gliwice: Silesian University of Technology Publishing House.

Garau, C., Desogus, G., Annunziata, A., Mighela, F. (2023). Mobility and health in the Smart City 3.0: Trends and innovations in Italian context. In *Smart cities and digital transformation: Empowering communities, limitless innovation, s*ustainable *d*evelopment and the *n*ext generation, London: Emerald, 105–127.

Gasco-Hernandez, M., Yerden, X., Burke, G. B., Gil-Garcia, J. R. (2022). The potential role of public libraries in a Quadruple Helix model of 'Smart City' development: Lessons from Chattanooga. *Sustainability*, 16, 1750.

Höglund, L., Linton, G. (2018). Smart specialization in regional innovation systems: A quadruple helix perspective. *R&D Management*, 48, 60–72.

Hussain, H. I., Haseeb, M., Kamarudin, F., Dacko-Pikiewicz, Z., Szczepańska-Woszczyna, K. (2021). The role of globalization, economic growth and natural resources on the ecological footprint in Thailand: Evidence from nonlinear causal estimations. *Processes*, 9(7), 1103.

Ivanova, I. (2014). Quadruple Helix systems and symmetry: A step towards helix innovation system classification. *Journal of Knowledge Economy*, 5, 357–369.

Izbash, A., Fomenko, O., Danylov, S. (2023). Innovative tools for implementing the smart city concept in architectural urbanism. *IP Conference Proceedings*, 2490(1), 030009.

Jonek-Kowalska, I. (2022). Municipal waste management in Polish cities: Is it really smart? *Smart Cities*, 5(4), 1635–1654.

Jonek-Kowalska, I. (2023). Assessing the effectiveness of air quality improvements in Polish cities aspiring to be sustainably smart. *Smart Cities*, 6(1), 510–530.

Jonek-Kowalska, I., Wolniak, R. (2023). *Smart cities in Poland: Towards sustainability and a better quality of life?* London: Routledge.

Jonek-Kowalska, I., Wolniak, R. (2021). Economic opportunities for creating smart cities in Poland: Does wealth matter? *Cities*, 114, 103222.

Jonek-Kowalska, I., Wolniak, R. (2021). The influence of local economic conditions on start-ups and local open innovation system. *Journal of Open Innovation: Technology, Market, and Complexity*, 7(2), 110.

Kalenyuk, I., Lukyanenko, L., Tsymbal, L., Stankevics, A., Uninets, I. (2023). The smart manufacturing: Imperatives and trends. *Financial and Credit Activity: Problems of Theory and Practice*, 5(52), 327–340.

Khan, T. S., Khan, N. U., Juneio, H. F. (2020). Smart city paradigm: Importance, characteristics, and implications. In *Advances in Science and Engineering Technology International Conferences*, 4 February–9 April 2020, Dubai, United Arab Emirates, 9118352.

Khavarian-Garmsir, A. R., Sharifi, A. (2022). Smart cities: Key definitions and new directions. In *Urban climate adaptation and mitigation*, London: Elsevier, 49–67.

Khurshudov, A. (2024). The smart city conundrum: Technology, privacy, and the quest for convenience. *Smart and Sustainable Built Environment*,. https://doi.org/10.1108/SASBE-12-2023-0377

Kinelski, G. (2022). Ewolucja koncepcji Smart City w aktywności zasobów miejskich – studium przypadku Górnośląsko-Zagłębiowskiej metropolii. *Przegląd Organizacji*, 2, 36–44.

Kinelski, G. (2022). Smart City 4.0 as a set of social synergies. *Polish Journal of Management Studies*, 26(1), 92–106.

Kondratenko, I., Mironov, D., Kulikova, E. (2023). Analysis of smart city energy efficiency technologies. *AIP Conference Proceedings*, 2812(1), 020079.

Konstantopoulou, E., Sklavos, N., Ognjanovic, I. (2024). Securing public safety mission-critical 5G communications of smart cities. *Signals and Communication Technology*, F1293, 61–74.

Kowalska, I. J., Wolniak, R. (2023). Sharing economies' initiatives in municipal authorities' perspective: Research evidence from Poland in the context of smart cities' development. *Sustainability*, 14(4), 2064.

Kramarz, M., Przybylska, E., Dohn, K., Jonek-Kowalska, I. (2022). *Urban logistics in a digital world: Smart cities and innovation.* London: Palgrave, 1–173.

Kulturalna Warszawa. (2024). https://polskieapps.pl/iphone-ipad/rozrywka/kulturalna-warszawa-aqaqmc.html (accessed 16.02.2024).

Kuzior, A., Kuzior, P. (2020). The quadruple helix model as a smart city design principle. *Virtual Economics*, 3(1), 39–57.

Kvalvik, P., Sánchez-Gordón, M., Colomo-Palacios, R. (2023). Beyond technology in smart cities: A multivocal literature review on data governance. *Aslib Journal of Information Management*, 75(6), 1235–1252.

Lee, J. H., Hancock, M. G., Hu, M. C. (2014). Towards an effective framework for building smart cities: Lessons from Seoul and San Francisco. *Technological Forecasting and Social Change*, 89, 80–90.

Li, D., Shang, X., Huang, G., Zhang, M., Feng, H. (2024). Can smart city construction enhance citizens' perception of safety? A case study of Nanjing, China. *Social Indicators Research*, 171, 937–965.

Locate Me. (2024). https://play.google.com/store/apps/details?id=com.anuraag.locateme&hl=pl&gl=US&pli=1 (accessed 16.02.2024).

Lokal. (2024). https://play.google.com/store/apps/details?id=com.mtdata.LokalWaterloo&hl=pl&gl=US (accessed 16.02.2024).

Lopes, C. T., Luciano, E. M. (2023). The role of the private sector on developing smart cities under a quadruple helix approach. *ACM International Conference Proceeding Series*, 470–472.

Lyu, K., Hao, M. (2020). Definition and history of smart cities: The development of cities and application of artificial intelligence technology in smart cities. In *AI-based services for smart cities and urban infrastructure*. Hershey, PA: IGI Global, 1–22.

Maddel, M., Gite, P., Sudhakar, K. N., Bakhare, R., Mittal, V. (2023). Development of specific mobile applications supporting the smart city trend. In *2nd International Conference on Edge Computing and Applications*, 19–21 July 2023, Namakkal, India, 195–202.

Madhee, K. H. (2024). Future of urban architectural design based on the concept of smart city. *Journal of Autonomous Intelligence*, 7(1), 1–10.

Makieła, Z. J., Stuss, M. M., Mucha-Ku´s, K., Kinelski, G., Budziński, M., Michałek, J. (2022). Smart City 4.0: Sustainable urban development in the metropolis GZM. *Sustainability*, 14, 3516.

Manfreda, A., Mijač, T. (2024). Smart city as a mix of technology, sustainability and well-being: A myth or reality? *IFIP Advances in Information and Communication Technology*, 699, 46–57.

Markard, J., Raven, R., Truffer, B. (2012). Sustainability transitions: An emerging field of research and its prospects. *Research Policy*, 41, 955–967.

McKenna, H. P. (2016). Is it all about awareness? People, Smart Cities 3.0, and evolving spaces for IT. In *2016 ACM SIGMIS Conference on Computers and People Research*, June 2–4, Virginia, USA, 47–56.

McKinsey Global Institute. (2018). Smart cities: Digital solutions for a more livable future. www.mckinsey.com/~/media/McKinsey/Industries/Public%20and%20Social%20Sector/Our%20Insights/Smart%20cities%20Digital%20solutions%20for%20a%20more%20livable%20future/MGI-Smart-Cities-Executive-summary.pdf (accessed 29-01-2024).

Mkrtychev, O., Starchyk, Y., Yusupova, S., Zaytceva, O. (2018). Analysis of various definitions for smart city concept. *IOP Conference Series: Materials Science and Engineering*, 365(2), 022065.

mObywatel. (2024). www.gov.pl/web/mobywatel (accessed 16.02.2024).

Neofotistos, M., Hanioti, N., Kefalonitou, E., Perouli, A. Z., Vorgias, K. E. (2023). A real-world scenario of citizens' motivation and engagement in urban waste management through a mobile application and smart city technology. *Circular Economy and Sustainability*, 3(1), 221–239.

Nextbike. (2024). https://nextbike.pl/dla-ciebie/ (accessed 16.02.2024).

Nordberg, K. (2015). Enabling regional growth in peripheral non-university regions: The impact of a Quadruple Helix intermediate organisation. *Journal of Knowledge Economy*, 6, 334–356.

Nugraha, Y. (2020). Building a smart city 4.0 ecosystem platform: An overview and case study. In *7th International Conference on ICT for Smart Society: AIoT for Smart Society*, 19–20 November 2020, Bandung, Indonesia, 930753.

Organisation for Economic Co-operation and Development. (2020). Better life initiative: Measuring well-being and progress. www.oecd.org/wise/better-life-initiative.htm (accessed 16.02.2024).

Osipova, M., Hornecker, E. (2023). Exploring the potential for smart city technology for women's safety. *ACM International Conference Proceeding Series*, 245–250.

Paskaleva, K., Evans, J., Watson, K. (2021). Co-producing smart cities: A Quadruple Helix approach to assessment. *European Urban and Regional Studies*, 28(4), 395–412.

Pellatt, J., Palfreman, J. (2023). Smart technology solution for a cleaner city: A case study of Dar es Salaam, Tanzania. *GeoJournal*, 88(5), 5221–5245.

Picioroaga, I.-I., Eremia, M., Sanduleac, M. (2018). Smart city: Definition and evaluation of key performance indicators. In *10th International Conference and Expositions on Electrical and Power Engineering*, 18–19 Oct. 2018, Iasi Romania, 8559763.

Piqué Huerta, J. M. (2019). *Understanding the urban development and the evolution of the ecosystems of innovation*. PhD thesis, Universidad Ramon Llull, Barcelona.

Qian, X. (2023). Evaluation on sustainable development of fire safety management policies in smart cities based on big data. *Mathematical Biosciences and Engineering*, 20(9), 17003–17017.

Raj, A., Shetty, S. D. (2024). Smart parking systems technologies, tools, and challenges for implementing in a smart city environment: A survey based on IoT & ML perspective. *International Journal of Machine Learning and Cybernetics*, 15(7), 1–22.

Rastog, R., Bhardwaj, M., Sharma, A. (2022). Crop and yield prediction through machine learning techniques to maximize production: 21st century sustainable approach for Smart Cities 5.0. In *4th International Conference on Advances in Computing, Communication Control and Networking*, 16–17 Dec. 2022, Greater Noida, India, 1286–1291.

Rastogi, R., Arora, P., Dhamija, L., Shrivastava, R. (2023). Intelligent online voting system for twenty-first century and Smart Cities 5.0: An empirical approach through blockchain with ML techniques. *Lecture Notes in Electrical Engineering*, 1056, 151–156.

Rele, M., Patil, D. (2023). Enhancing safety and security in renewable energy systems within smart cities. In *12th IEEE International Conference on Renewable Energy Research and Applications*, 105–114.

Ruiz-Vanoye, J. A. (2023). *Management, technology, and economic growth in smart and sustainable cities*,, 29 August–1 September 2023, Oshawa, Canada. https://doi.org/10.4018/979-8-3693-0373-3.

Sarjana, S. (2023). Smart city in supporting sustainable cities. In 2023 10th International Conference on Information Technology, Computer, and Electrical Engineering, 31 Aug.–1 Sep. 2023, Semarang, Indonesia, 365–370.

Schütz, F., Heidingsfelder, M., Schraudner, M. (2019). Co-shaping the future in Quadruple Helix innovation systems: Uncovering public preferences toward participatory research and innovation. *Journal of Design, Economics, and Innovation*, 5, 128–146.

Sharifi, A., Alizadeh, H. (2023). Societal smart city: Definition and principles for post-pandemic urban policy and practice. *Cities*, 134, 104207.

Smart City 4.0. (2022). www.skills4cities.eu/blog--news/smart-city-40 (accessed 16.02.2024).

Smart City 4.0. (2023). www.linkedin.com/pulse/smart-city-40-thinkco-pl/ (accessed 16.02.2024).

SOS Alarm. (2024). https://play.google.com/store/apps/details?id=se.sos.soslive&hl=pl&gl=US, (accessed 16.02.2024).

Stahl, B., Reiterer, A. (2022). Mobile mapping platform with integrated end-to-end data processing chain for smart city applications. *Proceedings of SPIE – The International Society for Optical Engineering*, 12269, 1226902.

Stecuła, K., Wolniak, R., Grebski, W. W. (2023). AI-driven urban energy solutions: From individuals to society: A review. *Energies*, 16(24), 7988.

Streitz, N. A., Riedmann-Streitz, C., Quintal, L. (2023). From 'smart-only' island towards lighthouse of research and innovation. *Lecture Notes in Computer Science*, 13325, 105–126.

Suzic, B., Ulmer, A., Schumacher, J. (2020). Complementarities and synergies of quadruple helix innovation design in smart city development. *2020 Smart Cities Symposium*, Prague, 25 June 2020, 9133961.

Svitek, M., Kozhevnikov, S. (2023). Smart City 5.0 as digital ecosystem of smart services: Basic concepts. In Smart *cities* and *digital transformation*: Empowering *communities, limitless innovation, sustainable development and the next generation*, Bingley: Emerald, 35–57.

Svitek, M., Kozhevnikov, S., Tencar, J., Bhattacharjee, S., Benes, V. (2023). Smart City 5.0 as the digital ecosystem of smart services: Practical applications. In *Smart cities and digital transformation: Empowering communities, limitless innovation,* sustainable development and the next generation, Bingley: Emerald, 327–356.

Svítek, M., Skobelev, P., Kozhevnikov, S. (2020). Smart City 5.0 as an urban ecosystem of smart services. *Studies in Computational Intelligence*, 853, 426–438.

Tamás, S. T., Dóra, S. (2023). Measuring the economic and environmental sustainability of cities with county rank, 2020–2021. *Teruleti Statisztika*, 63(1), 89–124.

Taratori, R., Rodriguez-Fiscal, P., Pacho, M. A., Koutra, S., Pareja-Eastaway, M., Thomas, D. (2021). Unveiling the evolution of innovation ecosystems: An analysis of triple, quadruple, and quintuple helix model innovation systems in European case studies. *Sustainability*, 13(14), 7582.

Teng, L. M., Yusoff, K. H., Mohammed, M. N., Sapari, N. M., Alfiras, M. (2024). Toward sustainable smart cities: Smart water quality monitoring system based on IoT technology. Studies in Systems, Decision and Control, 487, 577–593.

Toli, A. M., Murtagh, N. (2020). The concept of sustainability in smart city definitions. *Frontiers in Built Environment*, 6, 77.

Tomitsch, M., Ellison, A. (2023). A human-centred technology approach to pedestrian safety in smart cities. *Advances in Science, Technology and Innovation*, 19–32.

Trencher, G. (2019). Towards the Smart City 2.0: Empirical evidence of using smartness as a tool for tackling social challenges. *Technological Forecasting and Social Change*, 142, 117–128.

Van Den Berg, C., Verster, B. (2023). Advancing sustainable-smart innovations through a transdisciplinary learning intervention: Insights from the quintuple helix model. *European Conference on Innovation and Entrepreneurship*, 2, 883–890.

Vasudavan, H., Gunasekaran, S. S., Balakrishnan, S. (2019). Smart city: The state of the art, definitions, characteristics and dimensions. *Journal of Computational and Theoretical Nanoscience*, 16(8), 3525–3531.

Vătăşoiu, R.-I., Brătulescu, R.-A., Mitroi, S.-A., Mari-Anais S., Tudor, A.-M., Vintilă, A.-G. (2023). The importance of security and safety in a smart city. Smart Innovation, Systems and Technologies, 321, 11–23.

Vlahakis, V., Protopsaltis, A., Sarigiannidis, P. (2023). Smart city traffic monitoring using digital twin technology. *AIP Conference Proceedings*, 2909(1), 120014.

Wolniak, R. (2023). Smart mobility in smart city: Copenhagen and Barcelona comparison. *Scientific Papers of Silesian University of Technology*, 172, 679–697.

Wolniak, R., Grebski, W. (2023). Energy efficiency management in smart city: Smartphone applications aspects. *Scientific Papers of Silesian University of Technology*, 185, 605–615.

Wolniak, R. (2023). European Union smart mobility: Aspects connected with bike road system's extension and dissemination. *Smart Cities*, 6(2), 1009–1042.

Wolniak, R. (2023). Analysis of the bicycle roads system as an element of a smart mobility on the example of Poland provinces. *Smart Cities*, 6(1), 368–391.

Wolniak, R., Jonek-Kowalska, I. (2023). The creative services sector in Polish cities. *Journal of Open Innovation: Technology, Market, and Complexity*, 8(1), 17.

Wolniak, R., Gajdzik, B., Grebski, W. (2023). The implementation of Industry 4.0 concept in smart city. *Scientific Papers of Silesian University of Technology*, 178, 753–770.

Xu, H. (2023). Intelligent automobile auxiliary propagation system based on speech recognition and AI driven feature extraction techniques. *International Journal of Speech Technology*, 25(4), 893–905.

Yazici, A. M. (2023). The quintuple helix, industrial 5.0, and society 5.0. In Digitalization, sustainable development, and Industry 5.0: An organizational model for twin transitions, Bingley: Emerald, 317–336.

Yun, Y., Lee, M. (2020). Smart City 4.0 from the perspective of open innovation. *Journal of Open Innovation: Technology, Market, and Complexity*, 5(4), 92.

Zhou, Y. (2022). The application trend of digital finance and technological innovation in the development of green economy. *Journal of Environmental and Public Health*, 2022, 1064558.

2 Bright and dark sides of Smart City

2.1 The role of Smart City in improving the quality of urban life

Starting from the typology presented in Chapter 1 – six pillars of Smart City, in this chapter the benefits and problems and risks associated with the implementation of the Smart City concept will be presented in this arrangement. Thus, successively the benefits of Smart City implementation related to each pillar will be described, and in the next section we will present the problems and risks in each of the analysed areas.

The effective use of advanced technologies in **Smart Governance** contributes to the streamlining of administrative processes through automation, digitization and better integration of data. This, in turn, leads to more efficient public institutions (Pereira et al., 2018). Citizen participation increases through access to online platforms that allow residents to express opinions, raise issues and participate in decision-making processes (Vitálišová et al., 2022). Transparency of decision-making processes is increased through the provision of public information, online data and tools for tracking government actions.

Optimization of resources becomes possible through smart public management solutions, which translates into rational use of financial resources (Shafifi et al., 2024). Real-time data analysis supports accurate and timely public policy decisions. Integrated information systems facilitate the exchange of information between government sectors, eliminating unnecessary bureaucratic procedures.

Cyber security becomes a priority, protecting citizens' data and information systems from attacks. Improved communication between the administration and citizens reduces response times to reports and improves the overall quality of communication between the parties

DOI: 10.4324/9781003499992-3

(Sierdovski et al., 2022). As a result, smart governance contributes to better responsiveness of public administration, while increasing citizen participation and improving the quality of public services (Table 2.1).

The **Smart Economy** promotes sustainable development through the use of environmentally friendly technologies, efficient management of natural resources and support for businesses with low environmental impact (Youssef and Hajek, 2022). The development of the digital services sector, such as e-commerce, cloud services and fintech, contributes to the diversification of the economy and the creation of new jobs.

Smart cities support entrepreneurs by providing modern infrastructure, creating an innovation ecosystem and facilitating collaboration with R&D institutions. Improving access to global markets, assisting exports and enabling rapid response to changes in the business environment promotes business expansion (Wirsbinna, 2021). Additionally, the smart economy fosters labour market flexibility through support for remote work, digital communication tools and collaboration platforms. This enables companies to adapt to changing market conditions and meet employee preferences (Alam, 2023).

As a result, the introduction of Smart Cities in the smart economy supports dynamic economic development, fostering innovation, operational efficiency, sustainability and labour market flexibility (Table 2.2).

The implementation of Smart Cities in the area of **Smart Mobility** also brings with it a number of positive aspects (Table 2.3) that benefit the quality of life of residents and the efficiency of city operations (Qiu et al., 2023; Biswas et al., 2023). One of the key strengths in this regard is improving the accessibility and efficiency of public transportation. Through the use of advanced technologies, such as intelligent traffic management systems and real-time monitoring, public transportation routes and schedules can be optimized, resulting in reduced travel times and increased efficiency (Creutzig, 2020).

Smart cities also support the development of low-emission transport, such as electric vehicles and city bikes. This fosters reducing emissions of harmful substances, while improving air quality and limiting negative effects on the environment. Intelligent mobility systems also support the development of car-sharing and bike-sharing, which favours the sustainable development of individual transport. Innovations in the area of smart mobility contribute to improving road safety (Šurdonja et al., 2020). The use of advanced monitoring

Table 2.1 Positive aspects of Smart City implementation in the area of Smart Governance

Aspect	*Description*
Administrative efficiency	Smart City implementation in the area of smart governance contributes to the streamlining of administrative processes through automation, digitization and better integration of data, which increases the efficiency of public institutions.
Citizen participation	Smart cities enable better citizen participation through access to online platforms where residents can express their opinions, raise issues and participate in decision-making processes.
Process transparency	Smart governance contributes to greater transparency in decision-making processes, with access to public information, online data and tools to track government actions.
Optimization of resources	The introduction of smart solutions in the area of public management allows for better planning, resource allocation and optimization of public service delivery, which translates into rational use of financial resources.
Real-time data analysis	Access to real-time data and the ability to analyse it quickly supports more accurate and timely public policy decisions, which increases the efficiency of government and administration operations.
Integrated systems	Smart cities use integrated information systems that facilitate the exchange of information between different sectors of government, which promotes consistency of operations and eliminates unnecessary bureaucratic procedures.
Cyber security	The implementation of smart governance brings with it developed cyber security measures, ensuring that citizens' data and information systems are protected from attacks, which builds trust in electronic public services.
Communication improvement	The use of modern communication tools within the framework of smart governance facilitates interaction between the administration and citizens, which reduces response times to requests and improves the overall quality of communication between the parties.

Source: Authors' own work based on Vitálišová et al. (2022); McHenry (2013); Sierdovski et al. (2022); Shi and Cao (2022); Pereira et al. (2018); Smart Governance (2018); Sharifi et al. (2024); Amine (2024).

Table 2.2 Positive aspects of Smart City implementation in the area of Smart Economy

Aspect	*Description*
Economic innovations	The implementation of Smart Cities in the smart economy area stimulates innovation through the integration of modern technologies in business, which contributes to the development of the entrepreneurial sector and the creation of new business models.
Increased economic efficiency	Smart solutions enable optimization of production processes, resource management and logistics, which translates into increased economic efficiency and increased competitiveness of Smart City companies.
Sustainable development	The smart economy promotes sustainable development through the use of environmentally friendly technologies, efficient management of natural resources and support for low-impact businesses, helping to reduce the negative impact on the planet.
Development of the digital sector services	Smart cities foster the development of the digital services sector, which includes e-commerce platforms, cloud services, fintech or remote service delivery, among others, which contributes to diversifying the economy and creating new jobs.
Supporting entrepreneurship	Smart cities provide entrepreneurs with access to modern infrastructure, an innovation ecosystem and easier paths to collaborate with R&D institutions, which supports the growth of start-ups and enterprises.
Improved access to markets	The use of smart technologies facilitates access to global markets, aiding exports, as well as enabling companies to respond more quickly to changes in the business environment, which promotes business expansion at various levels.
Flexibility of the labour market	The smart economy fosters labour market flexibility by supporting remote work, digital communication tools and collaboration platforms to adapt to changing market conditions and employee preferences.

Source: Author's own work based on Zhong et al. (2024); Youssef and Hajek (2022); Wirsbinna (2021); Musamih et al. (2024); Alam (2023); Jasińska et al. (2023); Hepziba and Ebenesar (2023); Kaluarachchi (2022); Syalinda and Kusumatuti (2021); Hashemi (2022); Bonte (2022); Tyas et al. (2019); Jonek-Kowalska (2021).

Table 2.3 Positive aspects of Smart City implementation in the area of Smart Mobility

Aspect	*Description*
Smart public transport	The implementation of Smart Cities enables the development of smart public transportation, including smart stops, digital payments, real-time scheduling and route optimization, which improves the accessibility and efficiency of public transportation.
Development of electric transportation	Smart Cities promote the development of electric transportation, including electric buses, cars and bicycles. This approach contributes to reducing greenhouse gas emissions and improving air quality.
Car-sharing and bike-sharing systems	Smart Mobility supports the development of car-sharing and bike-sharing systems that allow efficient use of vehicles, reducing the number of private cars on the streets and contributing to sustainable development.
Smart parking lots	The implementation of smart parking management systems allows parking spaces to be found quickly and efficiently. This in turn reduces traffic jams, exhaust emissions and improves the user experience.
Communication between vehicles (v2v) and infrastructure (v2i)	Technologies that enable communication between vehicles and infrastructure improve road safety, optimize traffic flow and support emergency management.
Development of autonomous vehicles	Smart Cities are supporting the development of autonomous vehicles, which can improve traffic safety, efficiency and capacity.
Real-time traffic data	Smart cities collect real-time traffic data using sensors, cameras and other technologies. This information is used to optimize routes, manage congestion and improve overall mobility.
Innovations in logistics and delivery	The implementation of Smart Cities contributes to innovation in logistics and deliveries, enabling more efficient and sustainable delivery of goods in the urban area.

Source: Authors' own work based on Gironés and Vrščaj (2018); Benevolo et al. (2016); Šurdonja et al. (2020); Bolesnikov et al. (2023); Creutzog (2020); Nagy and Csiszar (2020); Sobnath et al. (2020); Celesti and Villari (2016).

systems, sensors and communication between vehicles enables quick response to potential threats and minimizes the risk of road accidents. Additionally, intelligent solutions support parking management, eliminating the problem of crowded streets and making it easier to find a parking space (Gironés and Vrščaj, 2018).

Data integration and analysis of information from various sources within smart mobility enables more effective decisions regarding spatial planning and urban infrastructure. This allows the city to be better adapted to the needs of its inhabitants, increasing the comfort of life and the efficiency of using available resources. As a result, the introduction of intelligent solutions in the area of smart mobility contributes to the creation of more sustainable, effective and resident-friendly cities, which is an important step towards the development of smart communities.

In turn, an important aspect of **Smart Environment** is supporting sustainable development through effective management of natural resources and minimizing emissions of harmful substances. Advanced technologies enable effective monitoring of air, water and soil quality, which allows for a quick response to potential environmental threats (Salleh et al., 2022). Energy efficiency increases thanks to intelligent energy management systems, optimization of energy consumption in buildings and support for energy production from renewable sources. Biodiversity conservation is promoted through greening urban spaces, creating urban gardens and monitoring and protecting animal breeding sites (Rui, 2017).

Smart cities engage in effective waste management, using monitoring systems, identifying sources of pollution and recycling technologies. This helps reduce waste and minimize the impact on the environment. In the area of transport, intelligent solutions help reduce pollutant emissions by optimizing vehicle traffic, supporting low-emission transport modes and promoting alternative means of transport. Introducing ecological education, access to information on ecology and sustainable development and initiating educational campaigns contribute to raising ecological awareness among residents (Kalauarachchi, 2021).

Smart cities also enable the integration of various environmental management systems, which allows for a comprehensive approach to ecological problems and effective environmental protection activities. As a result, the introduction of a smart environment contributes to the creation of more sustainable, ecological and resident-friendly cities (Table 2.4).

Table 2.4 Positive aspects of Smart City implementation in the area of Smart Environment

Aspect	*Description*
Sustainable development	The introduction of Smart Cities in the area of smart environment supports sustainable development through effective management of natural resources, minimizing emissions of harmful substances and supporting environmentally friendly practices.
Environmental monitoring	Advanced technologies enable effective monitoring of air, water and soil quality, which allows for quick response to potential environmental threats. Real-time data analysis enables informed decisions to be made to protect the urban ecosystem.
Energetic efficiency	A smart environment helps improve energy efficiency by using intelligent energy management systems, optimizing energy consumption in buildings and supporting the production of energy from renewable sources.
Biodiversity protection	Smart cities engage in the protection of biodiversity by greening urban spaces, creating urban gardens and monitoring and protecting animal breeding places. This works to balance urban ecosystems and improve environmental quality.
Effective waste management	Implementing a smart environment enables effective waste management through the use of monitoring systems, identification of pollution sources and implementation of recycling technologies. This helps reduce waste and minimize the impact on the environment.
Smart transportation systems	Intelligent transport solutions help reduce pollutant emissions by optimizing vehicle traffic, supporting low-emission transport modes and promoting alternative means of transport.
Ecological education	Smart environment encourages ecological education through access to information on ecology and sustainable development, initiating educational campaigns and promoting pro-ecological habits among residents.
Integrated management systems	The introduction of intelligent solutions in the area of environmental management enables the integration of various systems, which allows for a comprehensive approach to ecological problems and effective environmental protection activities.

Source: Authors' own work based on Amadeo et al. (2023); Salleh et al. (2022); Rui (2017); Janik et al. (2023); Kalaurachchi (2021); Sharifi et al. (2024); de Jong and Sun (2023); Ahmed (2023); Demirkesen et al. (2023); Jonek-Kowalska (2022); Jonek-Kowalska and Wolniak (2023).

In the area of **Smart Living**, an important element is increasing living comfort by automating home systems such as lighting, heating and air conditioning. These solutions foster adapting the home environment to the individual preferences of residents, which has a positive impact on their well-being (Sovacool and Furszyfer, 2020). The introduction of intelligent solutions promotes the energy efficiency of buildings, enabling monitoring and optimization of energy consumption. This not only translates into cost reduction, but also supports sustainable development by minimizing the impact on the environment (Wilson et al., 2017).

The safety of residents is increased thanks to advanced monitoring systems, security sensors and remote monitoring systems. These technologies enable a quick response to threats and increase the level of security in the place of residence (Diraco et al., 2023). In the area of health, smart living supports the provision of health services by enabling telemedicine, remote patient monitoring and access to health mobile applications. This contributes to improving health care by enabling health services to be delivered in a more accessible and efficient manner (Chang and Smith, 2023).

Smart cities facilitate communication between residents and public institutions, providing easier access to information, submitting reports and participating in participatory processes through modern communication platforms and tools (Demeritzi et al., 2023). The introduction of smart living also helps build communities through online platforms, social initiatives and applications, making it easier for residents to cooperate, exchange information, organize social events and build neighbourly relations. Smart solutions in the area of smart living contribute to improving the quality of life of residents, promoting a sustainable lifestyle, community integration and effective resource management (Table 2.5).

Smart cities enable access to innovative educational tools, e-learning and remote courses, supporting skills development and education. Through telemedicine, remote patient monitoring and health applications, Smart Cities support health care, contributing to improving the quality of medical services. In the community area, smart people foster community building through online platforms, social initiatives and applications, facilitating collaboration, information exchange and organizing social events. They also support the mobility of residents through intelligent transport systems, travel planning applications and transport sharing, which makes it easier to move around the city (Mun et al., 2022).

Table 2.5 Positive aspects of Smart City implementation in the area of Smart Living

Aspect	*Description*
Living comfort	The implementation of Smart Cities in the area of smart living contributes to increasing living comfort by automating home systems such as lighting, heating and air conditioning, adapting them to the preferences of residents.
Energy saving	Intelligent solutions support the energy efficiency of buildings by monitoring and optimizing energy consumption, which leads to cost reductions and reduced environmental impact.
Safety of residents	Smart living allows for improving the safety of residents thanks to advanced monitoring systems, security sensors, alarm systems and remote monitoring systems that provide protection against threats and incidents.
Health services	The introduction of intelligent solutions supports the provision of health services through telemedicine, remote patient monitoring and access to health mobile applications, which contributes to improving health care and health monitoring.
Adapted environment	Smart living allows for adapting the environment to the individual needs of residents, offering intelligent home systems that respond to preferences regarding lighting, temperature and entertainment, which contributes to the residents' better well-being.
Ease of communication	Smart cities support ease of communication between residents and public institutions, enabling access to information, submitting reports and participating in participatory processes through modern communication platforms and tools.
Community and integration	The introduction of smart living promotes community building through online platforms, social initiatives and applications that make it easier for residents to cooperate, exchange information, organize social events and build neighbourly relations.
Sustainable lifestyle	Smart solutions support a sustainable lifestyle by promoting low-emission transport, monitoring and optimizing energy consumption and promoting ecological habits, which contributes to achieving balance between people and the environment.

Source: Authors' work based on Diraco et al. (2023); Fico et al. (2017); Nikiforova et al. (2023); Demeritzi et al. (2023); Wilson et al. (2017); Sovacool and Furszyfer (2020); Kumar (2020); Chanag and Smith (2023); Wang and Zhou (2023); Chen and Chan (2023); Santosa et al. (2023); Miranda et al. (2024); Jonek-Kowalska and Wolniak (2023).

In the area of healthy lifestyles, **Smart People** support sustainable lifestyles by promoting low-emission transport, healthy lifestyles, environmental education and access to information about sustainable practices. They also introduce innovative initiatives that facilitate social integration between different groups, creating a more open and friendly society. As a result, Smart Cities in the area of smart people create conditions conducive to the development, health, community and sustainable lifestyle of residents, which contributes to a more attractive and effective functioning of societies.

From the above considerations, it is clear that the creation of Smart Cities comes with many undeniable advantages and benefits. They contribute to improving the quality of urban life and make the inevitable urbanization more friendly and less harmful. Nevertheless, amid the prevailing pride in Smart Cities, there is also a voice of criticism, which we treat in more detail in the next section.

2.2 Criticism of the Smart City concept on the literature on the subject

The introduction of **Smart Government** in the area of Smart Cities can face a number of challenges and negative aspects. One of the key risks is the issue of privacy and data protection. The development of smart public management systems requires the collection and processing of large amounts of data, which raises the risk of violating residents' privacy. Digital inequality poses another challenge, as exclusive access to modern technology can widen gaps between communities, introducing inequalities in access to the benefits of smart solutions. A significant divide in access to technology can lead to the exclusion of certain social groups (Vitálišová et al., 2022).

Cyber security is a significant threat, especially in the context of smart governance systems, where hacking attacks or data loss can lead to disruptions in city operations and pose potential risks to citizens. Lack of transparency and public participation in decision-making processes can breed public distrust. Citizens may feel excluded from decision-making processes, which can negatively affect the acceptance of implemented solutions (Pereira et al., 2018).

Excessive reliance on technology puts cities at risk of dependency on systems, which can lead to disruptions in public services if they fail or lack adequate technological competence. Costs and financing are also a major concern, as implementing smart government requires

Table 2.6 Positive aspects of Smart City implementation in the area of Smart People

Aspect	*Description*
Access to education	Introducing a Smart City in the area of smart people enables access to innovative educational tools, e-learning, remote courses and platforms, which supports the development of skills and education of residents.
Healthcare	Smart cities support healthcare through telemedicine, remote patient monitoring, access to health applications and innovative medical technologies, which contributes to improving the quality of healthcare.
Community support	Smart people promotes community building through online platforms, social initiatives and applications, making it easier for residents to cooperate, exchange information, organize social events and build neighbourly relations.
Mobility support	Smart cities support the mobility of residents through intelligent transport systems, journey planning applications, bike-sharing and electric cars, which make it easier to move around the city and reduce street congestion.
Citizen participation	Smart people enable better civic participation thanks to access to online platforms where residents can express their opinions, report problems and participate in decision-making processes regarding their environment.
Information accessibility	Smart cities provide easier access to public, cultural and social information, which increases residents' awareness of events and social initiatives, and facilitates the use of public and cultural services.
Sustainable lifestyle	Introducing smart people supports a sustainable lifestyle by promoting low-emission transport, healthy lifestyles, environmental education and access to information about sustainable practices, which contributes to improving the quality of life.
Easier social integration	Smart cities facilitate social inclusion through a variety of initiatives that foster mutual understanding, acceptance and cooperation between different social groups, creating a more open and friendly society.

Source: Authors' own work based on Mun et al. (2022); Smart Cities (2020); Putting (2020); Santosa et al. (2023); Chen and Chan (2023); Sharifi et al. (2024); Amine (2024).

significant investment, which can be a challenge for public budgets, especially in smaller cities. The lack of standardization in the area of smart government can hinder the integration of different systems and cooperation between different administrative sectors, which can lead to fragmentation and difficulties in the effective exchange of information (Table 2.7).

A significant problem in the smart economy is the increase in economic inequality, as modern technologies may not be available to all companies. Companies with smaller financial resources may find it difficult to adapt to modern technologies, leading to further concentration of resources in the hands of larger players (Youssef and Hajek, 2022). Another problem is the threat of job losses, especially in the sector of traditional industries that may be automated or replaced by modern technologies. Automation may lead to the need to retrain workers, which can be difficult and require additional investment in training and education.

Cyber security presents another challenge. As the complexity of smart economy systems increases, there is a greater risk of hacking attacks, data theft and disruption of digital platforms. This can affect confidence in modern solutions and hinder the full potential of the smart economy. Data privacy issues are also becoming important. As data is collected and analysed on a large scale, there is a risk of violations of individual privacy. Adequate data protection measures need to be put in place to avoid abuse and protect residents' privacy (Khan and Khan, 2023).

The introduction of a smart economy may also run the risk of becoming dependent on technology and focusing on short-term gains at the expense of long-term sustainability. An imbalance between technological advances and socioeconomic aspects of development can lead to structural problems and lack of sustainability. Therefore, it is important to approach the implementation of the smart economy with caution, taking into account social, economic and ethical aspects, in order to minimize negative impacts and evenly share the benefits of modern technologies (Table 2.8).

An important issue in smart mobility is the potential inequality of access to modern transportation. Individuals or groups with fewer financial resources or limited access to technology may find it difficult to access innovative mobility services, increasing potential social inequality. The introduction of modern means of transportation, such as electric or autonomous vehicles, may involve risks to traffic safety.

Table 2.7 Negative aspects and problems of Smart City implementation in the area of Smart Governance

Aspect	*Description*
Privacy and data protection	Smart City implementation in the area of smart governance carries the risk of violating residents' privacy and mismanaging personal data. It is necessary to provide adequate safeguards to avoid unauthorized access or data processing.
Digital inequalities	Exclusive access to modern technologies in the area of smart governance can lead to the accumulation of digital inequalities when not all residents have equal access to the benefits of smart solutions. Social and economic aspects need to be taken into account.
Security risks	Smart governance carries cyber security risks, such as hacking attacks on information systems, which can lead to data loss, disruption of city operations or potential threats to citizens. Effective security measures are needed.
Lack of transparency and participation	Implementing smart governance requires ensuring transparency and citizen participation in decision-making processes. Lack of openness can lead to public distrust and lack of acceptance of the introduced solutions, which can hinder the effective functioning of the Smart City.
Dependence on technology	Overdependence on technology can lead to a situation where system failures or lack of adequate technological competence can result in disruptions to public services and administrative processes in the city.
Costs and financing	Implementing smart governance requires significant financial investment, which can be a challenge for public budgets. It is necessary to focus on a sustainable funding model and to consider the costs of maintaining and updating systems in the long run.
Lack of standardization	Lack of standardization in the area of smart governance can lead to difficulties in integrating different systems and hinder cooperation between different administrative sectors. It is necessary to introduce standards to enable effective information exchange and cooperation between different institutions.

Source: Authors' own work based on Vitálišová et al. (2022); Garau et al. (2017); Cirella et al. (2023); Periñán-Pascual, 2023; Pereira et al. (2018); Leonhardt et al. (2023); Ozdamli and Nawalia (2023); Wielicka-Gańczarczyk and Jonek-Kowalska (2023); Wolniak and Jonek-Kowalska (2023).

Table 2.8 Negative aspects and problems of Smart City implementation in the area of Smart Economy

Aspect	*Description*
Economic inequalities	Smart City implementation in the smart economy area can generate economic inequality when not all entrepreneurs or residents have equal access to modern technologies. This can lead to deepening social and economic divisions in the city.
Job security	Automation and technology development can lead to job losses in some sectors of the economy, especially in traditional industries. It is necessary to adapt to the changes and provide adequate support measures and retraining for workers to avoid structural unemployment.
Dependence on large corporations	The introduction of the smart economy may result in increased dependence on large corporations, which have greater financial and technological resources. This could lead to reduced competition, difficulties for smaller companies, and result in a small number of players dominating the market.
Cyber risk	The development of the smart economy brings with it cyber risks associated with hacking attacks on financial, transaction systems and other economy-related platforms. Adequate protection against cyber threats is crucial to maintaining stability and security in the smart economy.
Lack of standards and interoperability	The lack of standardization and interoperability in smart economy systems can lead to difficulties in exchanging information between different platforms, making it difficult for companies and institutions to work together. Standards are needed to enable seamless integration of different systems and improve operational efficiency.
Loss of privacy	The development of the smart economy may lead to a loss of privacy due to intensive data analysis, monitoring of consumer behaviour and collection of customer information. Adequate regulations and privacy protection mechanisms are needed to minimize the risk of misuse of personal data.
Regulatory uncertainty	The dynamic development of the smart economy can lead to regulatory uncertainty when existing regulations fail to keep up with the pace of technological change. Regulations need to be adjusted to ensure a balance between the development of innovative solutions and the protection of public interests.

Source: Authors' own work based on Youssef and Hajek (2022); Amine (2024); Cirella et al. (2023); Marchesani (2023); Agboola and Findikgil (2023); Khan and Khan (2023).

Technical failures, software bugs and hacking attacks pose potential risks to drivers, pedestrians and other traffic participants (Gironés and Vrščaj, 2018; Lytras, 2023).

Also, urban infrastructure congestion is one of the challenges in the area of smart mobility. The introduction of modern solutions, such as electric cars and car-sharing systems, can lead to congestion in streets, parking lots and other urban spaces, especially in cities with high population density. The issue of battery range and durability in electric vehicles is becoming important, as limited range and battery recycling issues can pose challenges to smart mobility sustainability (Goumiri et al., 2023).

Additionally, the lack of standardization and interoperability between different mobility systems can lead to hindered user experience and limit the full potential of smart mobility. Violations of privacy and security of personal data are also a concern in the context of collecting data on traffic and travel habits (Apata et al., 2023). Effective privacy measures need to be put in place to avoid abuse and threats to residents' privacy. The introduction of new technologies and automation in the area of smart mobility may also lead to the loss of jobs in traditional transportation sectors, which requires simultaneous consideration of re-education and retraining of employees (Table 2.9).

In terms of the smart environment, one significant challenge is the generation of new sources of electronic pollution. High production of electronic devices, sensors and other components can lead to electronic waste management problems, requiring effective recycling and disposal strategies. Developing a smart environment can involve the loss of natural habitats and green spaces. The construction of new facilities, installation of sensors or infrastructure for modern systems can negatively affect local ecosystems, requiring simultaneous measures for nature conservation and sustainable urban planning (Krivykh, 2020).

Implementation of a smart environment often requires intensive consumption of natural resources at the stage of production, transportation and installation of new technologies. The manufacture of electronic devices, batteries or sensors can generate greenhouse gas emissions and use natural resources, challenging the long-term sustainability of Smart City projects. The introduction of smart technologies in the smart environment area can lead to increased demand for electricity. Devices, sensors, communication networks and other

Table 2.9 Negative aspects and problems of Smart City implementation in the area of Smart Mobility

Aspect	*Description*
Access inequalities	The introduction of smart mobility may lead to an increase in inequality in accessibility to modern means of transportation. Some segments of society may find it difficult to access modern mobility services due to financial, technological or geographic constraints, increasing potential social inequalities.
Infrastructure congestion	The development of smart mobility can lead to congestion of urban infrastructure, especially if modern modes of transportation, such as electric vehicles and urban bicycles, become more popular. Failure to prepare infrastructure to handle the large number of new vehicles can lead to congestion in streets, parking lots and other urban spaces.
Threats to traffic safety	The introduction of modern mobility technologies, such as autonomous vehicles, carries potential traffic safety risks. Technical failures, software bugs or hacking attacks can lead to dangerous situations for drivers, pedestrians and other traffic participants, requiring effective safety measures.
The problem with battery range and durability	The introduction of electric vehicles as part of smart mobility brings challenges related to battery range and longevity. The limited range of electric vehicles and the issue of battery recycling can pose challenges to sustainability. It is necessary to develop efficient and environmentally friendly power solutions for electric vehicles.
The problem of system interoper-ability	The introduction of a variety of smart mobility systems can lead to interoperability problems. The lack of standards or harmonized solutions can make it difficult for different modes of transportation, payment systems or platforms to work together, which can limit the full potential of smart mobility.
Threats to privacy and personal data	The development of smart mobility may lead to privacy and personal data security breaches, especially in the context of collecting data on traffic and travel habits. Effective privacy measures need to be put in place to avoid abuse and threats to residents' privacy.
Job losses in traditional sectors	Automation and the development of smart mobility may lead to job losses in traditional transportation sectors, such as drivers and vehicle service station employees. The introduction of modern solutions requires simultaneous consideration of re-education and retraining of workers to minimize the social impact of job losses.

Source: Authors' own work based on Mounce et al. (2020); Raj and Shetty (2024); Apata et al. (2023); Lytras (2023); Kuo et al. (2023); D'alberto and Giudici (2023); Goumiri et al. (2023); Gironés and Vrščaj (2018).

components of Smart City systems require power, which can result in increased energy consumption (Sokołowski and Visvizi, 2023).

The implementation of modern smart environment technologies involves cyber security risks. Hacking attacks can lead to disruption of Smart City systems, compromise of environmental data and potential threats to residents. The introduction of smart environment monitoring and data collection systems can raise privacy and surveillance concerns. Sensors monitoring the environment can record data on residents' activities, which requires meticulous privacy protection and clear regulations.

It is worth noting that the deployment of diverse systems in the smart environment area can lead to problems with integration and cooperation between them. Lack of standardization and harmonized solutions can hinder the effective combination of data and functions of different technologies, which limits the full potential of smart environment solutions (Table 2.10).

A key challenge in smart living is related to invasion of residents' privacy. Systems for monitoring and analysing daily life data can lead to the collection and storage of detailed information about residents' habits, preferences or movements, which can raise privacy concerns. Implementing smart living solutions can also result in dependence on technology. Residents who become overly dependent on modern systems may experience difficulties with emergencies, power loss, or system disruptions, affecting daily life (Wang and Zhou, 2023).

Social inequality and digital exclusion are another challenge. The elderly, technologically uninformed or those with lower incomes may face difficulties in using modern technologies, leading to inequalities in access to the benefits of smart living. Excessive monitoring, which is inherent to smart living, can lead to feelings of surveillance and loss of privacy for residents. Constant observation and analysis of daily life data can raise concerns about individual freedom and lead to conflicts with privacy values (Samuel et al., 2022).

Increased consumerism is another issue that may arise from the implementation of smart living. The incentive to purchase more smart devices or gadgets can result in increased electronic waste, consumption of natural resources and negative environmental impacts. Moreover, excessive use of smart living systems can lead to mental health risks for residents. Dependence on technology, constant access to information or constant monitoring can affect stress levels, concentration and overall well-being. Effectively managing these challenges requires a holistic approach that considers the balance between the

Table 2.10 Negative aspects and problems of Smart City implementation in the area of Smart Environment

Aspect	*Description*
Electronic pollution	Smart City implementation in the smart environment area is generating new sources of pollution in the form of electronic waste. High production of electronic devices, sensors and other components can lead to electronic waste management problems, requiring effective recycling and disposal strategies to minimize environmental impact.
Loss of natural habitats and green spaces	Developing a smart environment often involves investment in technology infrastructure, which can lead to the loss of natural habitats and green spaces. The construction of new facilities, installation of sensors or infrastructure for modern systems can negatively affect local ecosystems, requiring simultaneous measures for nature conservation and sustainable urban planning.
Consumption of natural resources	Implementation of a smart environment often requires intensive consumption of natural resources at the stage of production, transportation and installation of new technologies. The manufacture of electronic devices, batteries or sensors can generate greenhouse gas emissions and use natural resources, challenging the long-term sustainability of Smart City projects.
Increased energy demand	The introduction of smart technologies in the smart environment area can lead to increased demand for electricity. Devices, sensors, communication networks and other components of Smart City systems require power, which can result in increased energy consumption. To minimize the negative impact, it is necessary to use efficient energy solutions and promote renewable energy sources.
Cyber security issues	The introduction of modern technologies in the smart environment area involves cyber security risks. Hacking attacks can lead to disruption of Smart City systems, compromise of environmental data and potential threats to residents. It is necessary to implement effective cyber security protection, monitoring and management measures to minimize these risks.

Table 2.10 (Continued)

Aspect	*Description*
Privacy and surveillance concerns	Implementing monitoring and data collection systems in the smart environment can raise privacy and surveillance concerns. Sensors monitoring the environment can record data on residents' activities, which requires meticulous privacy protection and clear regulations. It is necessary to include ethical standards and transparency in the collection and processing of environmental data.
Problems with system integration	The introduction of diverse systems in the smart environment area can lead to problems of integration and cooperation between them. Lack of standardization and harmonized solutions can make it difficult to effectively combine the data and functions of different technologies, limiting the full potential of smart environmental solutions.

Source: Authors' own work based on Lam et al. (2022); Amadeo et al. (2023); Qian, (2023); Sharifi et al. (2024); Sokołowski and Visvizi (2023); Magloacani (2023); Cirella et al. (2023); Erces et al. (2023); Osipova et al. (2023).

benefits of smart solutions and the privacy, social equity and health of residents (Table 2.11).

The issue of educational inequality is also important for the smart people area in Smart Cities. Residents with limited access to modern technology or low levels of digital literacy may have difficulty using smart solutions, widening the gap between social groups in terms of education and technological skills. Implementing smart solutions in the area of smart people can also result in dependence on technology. Residents who become heavily dependent on smart devices or systems may experience difficulties with emergencies, power loss or system disruptions, affecting daily life (Vitálišová et al., 2022).

Social inequality and digital exclusion are another challenge. People with lower incomes or living in areas with weaker technological infrastructure may be excluded from using modern solutions, leading to socio-technological inequalities. The integration of technology into residents' daily lives carries privacy risks. Data collected by smart systems can be used for profiling or monitoring, raising concerns about excessive intrusion into individual privacy and potential data abuse (Gurashi, 2019).

Table 2.11 Negative aspects and problems of Smart City implementation in the area of Smart Living

Aspect	*Description*
Privacy violations and monitoring	The introduction of smart living, which includes monitoring and analysis of data related to residents' daily lives, can lead to privacy violations. Monitoring systems can collect information about residents' habits, preferences and behaviours, which requires meticulous privacy protection and the application of appropriate regulations to avoid abuse or unauthorized access to data.
Dependence on technology	Dependence of smart living on modern technology can result in problems in the event of system failures, power loss or disruptions. Residents who become too dependent on smart solutions can experience difficulties when these systems do not work as expected, requiring a balanced approach to introducing new technologies.
Digital exclusion and social inequality	The implementation of smart living can lead to digital exclusion, especially among the elderly, the technologically ignorant or those with lower incomes. Inequalities in access to and ability to use modern technologies can cause not all residents to benefit equally from smart solutions, increasing existing social inequalities.
Problems associated with excessive monitoring	Excessive monitoring in smart living can lead to feelings of surveillance and loss of privacy for residents. Constant observation and analysis of daily life data can raise concerns about individual freedom and lead to conflicts with privacy values. Appropriate safeguards and monitoring restrictions are needed to minimize these negative effects.
Increase in consumerism	The implementation of smart living can affect the growth of consumerism, encouraging the purchase of more smart devices or gadgets. This excessive consumerism can result in increased electronic waste, consumption of natural resources and negative environmental impacts. A balanced approach to the use of modern technology is needed to avoid an uncontrolled increase in consumerism.
Threats to mental health	Excessive use of smart living systems, especially in the context of constant network connectivity or instant access to information, can lead to mental health problems. Dependence on technology, constant access to information or constant monitoring can affect residents' stress levels, concentration and overall well-being.

Source: Authors' own work based on Bianchini et al. (2018); Mańka-Szulik et al. (2023); Chang and Smith (2023); Wang and Zhou (2023); Chen and Chan (2023); Wolniak and Jonek-Kowalska (2021); Jonek-Kowalska (2023).

The introduction of Smart Cities without proper consideration of public participation may lead to a lack of active involvement of residents. Decisions on technological innovation and community development that are not consulted with residents can create dissatisfaction and a lack of a sense of participation in the decision-making process. Automation and Smart City development can lead to a loss of traditional skills, especially in the labour sector. Residents whose skills are tied to traditional forms of work may experience job losses or the need for retraining, requiring appropriate education and support measures for workers (Shafifi et al., 2024).

The introduction of Smart Cities can contribute to widening inequalities in access to technology. People with lower incomes or living in areas with weaker technological infrastructure may be excluded from using modern solutions, leading to socio-technological inequalities. Introducing Smart Cities without proper consideration of social values can lead to an imbalance between technology and community needs. Decisions based solely on technological advances, without considering ethical, cultural or social aspects, can lead to conflicts and non-acceptance of new solutions (Table 2.12).

Table 2.12 Negative aspects and problems of Smart City implementation in the area of Smart People

Aspect	*Description*
Education inequalities	Implementing Smart Cities in the area of smart people can exacerbate educational inequality. Residents with limited access to modern technology or low levels of digital literacy may have difficulty using smart solutions, widening the gap between social groups in terms of education and technological skills.
Technology access inequalities	The implementation of Smart Cities can contribute to widening inequalities in access to technology. People with lower incomes or living in areas with weaker technological infrastructure may be excluded from using modern solutions, leading to socio-technological inequalities.
Privacy risks	The integration of technology into residents' daily lives carries privacy risks. Data collected by smart systems can be used for profiling or monitoring, raising concerns about excessive intrusion into individual privacy and potential data abuse.

(*Continued*)

Table 2.12 (Continued)

Aspect	*Description*
Lack of public participation	The introduction of Smart Cities without proper consideration of public participation may lead to a lack of active involvement of residents. Decisions on technological innovation and community development that are not consulted with residents can create dissatisfaction and a lack of a sense of participation in the decision-making process.
Loss of traditional skills	Automation and Smart City development can lead to a loss of traditional skills, especially in the labour sector. Residents whose skills are tied to traditional forms of work may experience job losses or the need for retraining, requiring appropriate education and support measures for workers.
Imbalance between technology and social values	Introducing Smart Cities without proper consideration of social values can lead to an imbalance between technology and community needs. Decisions based solely on technological advances, without considering ethical, cultural or social aspects, can lead to conflicts and non-acceptance of new solutions.

Source: Authors' own work based on Gurashi et al. (2019); Mun et al. (2022); Amine (2024); Sharifi et al. (2024); Vitálišová et al. (2022); Vătăşoiu et al. (2023); Jonek-Kowalska (2022).

The considerations presented in this subsection signal preliminary problems and difficulties associated with the creation and development of Smart Cities. They are the seeds of possible exclusions of a diverse nature, which will be described in more detail in subsequent chapters of the monograph.

Bibliography

Agboola, O. P., Findikgil, M. M. (2023). A comparative framework analysis of the strategies, challenges and opportunities for sustainable smart cities, *Fostering Sustainable Development in the Age of Technologies*, 187–211.

Ahmed, A. (2023). Smart cities and IoT: Examining the potential benefits and challenges of using IoT to create more efficient and sustainable urban environments. *International Journal of Intelligent Systems and Applications in Engineering*, 11(10s), 306–320.

Alam, M. H. (2023). Role of the governance and good governance to build a smart economic and smart city: A case study of Bangladesh. *Technology and Talent Strategies for Sustainable Smart Cities: Digital Futures*, 103–115.

Amadeo, M., Campolo, C., Ruggeri, G., Molinaro, A. (2023). Edge caching in IoT smart environments: Benefits, challenges, and research perspectives toward 6G. *Internet of Things*, 53–73.

Amine, R. (2024). Smart cities: Development and benefits. In Belaïd, F., Arora, A. (eds), *Smart cities*. Studies in Energy, Resource and Environmental Economics. Cham: Springer. https://doi.org/10.1007/978-3-031-35664-3_4.

Apata, O., Bokoro, P. N., Sharma, G. (2023). The risks and challenges of electric vehicle integration into smart cities. *Energies*, 16(14), 5274.

Benevolo, C., Dameri, R. P., D'Auria, B. (2016). Smart mobility in smart city action taxonomy, ICT intensity and public benefits. *Lecture Notes in Information Systems and Organisation*, 11, 13–28.

Bianchini, D., De Antonellis, V., Melchiori, M., Bellagente, P., Rinaldi, S. (2018). Data management challenges for smart living. *Lecture Notes of the Institute for Computer Sciences, Social-Informatics and Telecommunications Engineering*, 189, 131–137.

Biswas, S., Yao, Z., Yan, L., Asiri, F., Masud, M. (2023). Interoperability benefits and challenges in smart city services: Blockchain as a solution. *Electronics*, 12(4), 1036.

Bolesnikov, M., Dumnić, B., Popadić, B., Stijačić, M. P., Radišić, M. (2023). Benefits of the smart mobility as an integral part of smart city Novi Sad. In *XXIX Skup Trendovi Razvoja: Univerzitet Pred Novim Izazovima*, Vrnjačka Banja, Serbia. www.researchgate.net/publication/368576004_BENEFITS_OF_THE_SMART_MOBILITY_AS_AN_INTEGRAL_PART_OF_SMART_CITY_NOVI_SAD (accessed 19.02.2024).

Bonte, D. (2022). Role of smart cities for economic development. www.ashb.com/wp-content/uploads/2020/04/IS-2018-215.pdf (accessed 19.02.2024).

Celesti, A., Villari, M. (2016). FrontierCities: Leveraging FIWARE for advantages in smart mobility. *Communications in Computer and Information Science*, 567, 450–451.

Chang, S., Smith, M. K. (2023). Residents' quality of life in smart cities: A systematic literature review. *Land*, 12(4), 876.

Chen, Z., Chan, I. C. C. (2023). Smart cities and quality of life: A quantitative analysis of citizens' support for smart city development. *Information Technology and People*, 36(1), 263–285.

Cirella, G. T., Domańska, A., Orobello, C. (2023). Creating smart cities in Poland: Opportunities, obstacles, and the missing link. *Lecture Notes in Networks and Systems*, 808, 14–25.

Creutzig, F. (2020). Leveraging the benefits of smart mobility via an integrated data platform. https://il.boell.org/sites/default/files/2020-11/Leveraging-the-Benefits-of-Smart-Mobility-via-an-Integrated-Data-Platform.pdf (accessed 19.02.2024).

D'Alberto, R., Giudici, H. (2023). A sustainable smart mobility? Opportunities and challenges from a big data use perspective. *Sustainable Futures*, 6, 100118.

de Jong, M., Sun, L. (2023). Eco and low-carbon, smart and sponge: Potential and delusion in realizing environmental benefits from sustainable city branding. *Handbook on China's urban environmental governance*, London: Edward Elgar, 186–200.

Demertzi, V., Demertzis, S., Demertzis, K. (2023). An overview of cyber threats, attacks and countermeasures on the primary domains of smart cities. *Applied Sciences*, 13(2), 790.

Demirkesen, S., Zhang, C., Kumar, D. (2023). Integrating BIM and GIS for disaster management in smart cities: Key benefits and challenges. *ASCE Inspire 2023: Infrastructure Innovation and Adaptation for a Sustainable and Resilient World*, 16–18 November 2023, Arlington, VA, 330–338.

Diraco, G., Rescio, G., Siciliano, P., Leone, A. (2023). Review on human action recognition in smart living: Sensing technology, multimodality, real-time processing, interoperability, and resource-constrained processing. *Sensors*, 23(11), 5281.

Elhofy, M., Abdelaziz, M., Omran, I., Abdelwahab, M. (2023). Effects of independent wheels steering system on vehicle cornering performance and road safety of the smart cities. *Ain Shams Engineering Journal*, 14(6), 102097.

Érces, G., Rácz, S., Vass, G., Varga, F. (2023). Fire safety in smart cities in Hungary with regard to urban planning. *Journal of Integrated Disaster Risk Management*, 13(2), 104–128.

Fico, G., Montalva, J., Medrano, A., Cea, G., Arredondo, M. T. (2017). Co-creating with consumers and stakeholders to understand the benefit of internet of things in smart living environments for ageing well: The approach adopted in the Madrid deployment site of the activage large scale pilot. *IFMBE Proceedings*, 65, 1089–1092.

Garau, C., Balletto, G., Mundula, L. (2017). A critical reflection on smart governance in Italy: Definition and challenges for a sustainable urban regeneration. *Smart and Sustainable Planning for Cities and Regions*, 235–250.

Gironés, E. S., Vrščaj, D. (2018). Who benefits from smart mobility policies? The social construction of winners and losers in the connected bikes projects in the Netherlands. *Governance of the Smart Mobility Transition*, 85–101.

Goumiri, S., Yahiaoui, S., Djahel, S. (2023). Smart mobility in smart cities: Emerging challenges, recent advances and future directions. *Journal of Intelligent Transportation Systems: Technology, Planning, and Operations*, 1–37. https://doi.org/10.1080/15472450.2023.2245750

Gurashi, R. (2019). Smart people and prosumers: The individual challenge to the fourth industrial revolution. *Smart Society: A Sociological Perspective on Smart Living*, 30–47.

Hashemi, A. A. A. A. (2022). Smart cities and their impact on economic sustainability: A contemporary view within the framework of concepts and experiences. *Sociology International Journal*, 6(6), 364–370.

Hepziba Gnanamalar, R., Ebenesar Anna Bagyam, J. (2023). Eco-friendly blockchain for smart cities. *Green Blockchain Technology for Sustainable Smart Cities*, 65–96.

Janik, A., Ryszko, A. Szafraniec, M. (2023). Intelligent and environmentally friendly solutions in smart cities' development: Empirical evidence from Poland. *Smart Cities*, 6(2), 1202–1226.

Jasińska, E., Preweda, E., Łazarz, P. (2023). Renewable energy sources in the residential property market, exemplified by the city of Krakow (Poland). *Sustainability*, 15(10), 7743.

Jonek-Kowalska, I. (2022). Health care in cities perceived as smart in the context of population aging: A record from Poland. *Smart Cities*, 5(4), 1267–1292.

Jonek-Kowalska, I. (2022). Housing infrastructure as a determinant of quality of life in selected Polish smart cities. *Smart Cities*, 5(3), 924–946.

Jonek-Kowalska, I. (2022). Municipal waste management in Polish cities: Is it really smart? *Smart Cities*, 5(4), 1635–1654.

Jonek-Kowalska, I. (2023). The exclusiveness of smart cities: Myth or reality? Comparative analysis of selected economic and demographic conditions of Polish cities. *Smart Cities*, 6(5), 2722–2741.

Jonek-Kowalska, I., Wolniak, R. (2023). *Smart cities in Poland: Towards sustainability and a better quality of life?* London: Routledge.

Jonek-Kowalska, I., Wolniak, R. (2021). Economic opportunities for creating smart cities in Poland: Does wealth matter? *Cities*, 114, 103222.

Kalauarachchi, Y. (2021). Potential advantages in combining smart and green infrastructure over silo approaches for future cities. *Frontiers of Engineering Management*, 8(1), 98–108.

Kaluarachchi, Y. (2022). Implementing data-driven smart city applications for future cities. *Smart Cities*, 5(2), 455–474.

Khan, R. A., Khan, M. W. (2023). Cyber security's influence on smart cities: Challenges and solutions. *AIP Conference Proceedings*, 2821(1), 040033.

Krivykh, E. G. (2020). Smart environment: Problems of social identity. *IOP Conference Series: Materials Science and Engineering*, 775(1), 012023.

Kumar, T. M. (2020). *Smart living for smart cities community study, ways and means*. Berlin: Springer.

Kuo, Y.-H., Leung, J. M. Y., Yan, Y. (2023). Challenges and vulnerability evaluation of smart cities in IoT device based on cybersecurity mechanism. *Expert Systems*, 40(4), e13113.

Lam, L., Fadrique, L., Bin Noon, G., Shah, A., Morita, P. P. (2022). Evaluating challenges and adoption factors for active assisted living smart environments. *Frontiers in Digital Health*, 4, 891634.

Leonhardt, S., Nusser, T., Görres, J., Stryi-Hipp, G., Eckhard, M. (2023). *Innovations and challenges of the energy transition in smart city districts*, Berlin: De Gruvter.

Lytras, M. D. (2023). Future smart cities research: Identifying the next generation challenges. In *Smart cities and digital transformation: Empowering communities, limitless innovation, sustainable development and the next generation*, Bingley: Emerald, 1–11.

Magliacani, M. (2023). How the sustainable development goals challenge public management: Action research on the cultural heritage of an Italian smart city. *Journal of Management and Governance*, 27(3), 987–1015.

Mańka-Szulik, M., Krawczyk, D., Wodarski, K. (2023). Residents' perceptions of challenges related to implementation of smart city solutions by local government. *Sustainability*, 15(11), 8532.

Marchesani, F. (2023). *The global smart city: Challenges and opportunities in the digital age*, Bingley: Emerald.

McHenry, M. P. (2013). Technical and governance considerations for advanced metering infrastructure/smart meters: Technology, security, uncertainty, costs, benefits, and risks. *Energy Policy*, 59, 834–842.

Miranda, R., Alves, C., Sousa, R., Novais, P., Machado, J. (2024). Revolutionising the quality of life: The role of real-time sensing in smart cities. *Electronics*, 13(3), 550.

Mounce, R., Beecroft, M., Nelson, J. D. (2020). On the role of frameworks and smart mobility in addressing the rural mobility problem. *Research in Transportation Economics*, 83, 100956.

Mun Chye, C., Fahmy-Abdullah, M., Sufahani, S. F., Bin Ali, M. K. (2022). A study of smart people toward smart cities development. In Kaiser, M. S., Ray, K., Bandyopadhyay, A., Jacob, K., Long, K. S. (eds), *Proceedings of the third international conference on trends in computational and cognitive engineering.* Lecture Notes in Networks and Systems, vol. 348. Singapore: Springer. https://doi.org/10.1007/978-981-16-7597-3_21.

Musamih, A., Dirir, A., Yaqoob, I., Jayaraman, R., Puthal, D. (2024). NFTs in smart cities: Vision, applications, and challenges. *IEEE Consumer Electronics Magazine*, 13(2), 9–23.

Nagy, S., Csiszár, C. (2020). The quality of smart mobility: A systematic review. *Scientific Journal of Silesian University of Technology: Series Transport*, 109, 117–127.

Nikiforova, A., Flores, M. A. A., Lytras, M. D. (2023). The role of open data in transforming the society to society 5.0: A resource or a tool for SDG-compliant smart living? In *Smart cities and digital transformation: Empowering communities, limitless innovation, sustainable development and the next generation*, Bingley: Emerald, 219–252.

Osipova, M., Hornecker, E. (2023). Exploring the potential for smart city technology for women's safety. *ACM International Conference Proceeding Series*, 245–250.

Ozdamli, F., Nawaila, M. B. (2023). Analysing the challenges and opportunities of smart cities. *Signals and Communication Technology*, F1293, 93–111.

Pereira, G. V., Parycek, P., Falco, E., Kleinhans, R. (2018). Smart governance in the context of smart cities: A literature review. *Information Policy*, 23(2), 143–162.

Periñán-Pascual, C. (2023). From smart city to smart society: A quality-of-life ontological model for problem detection from user-generated content. *Applied Ontology*, 18(3), 263–306.

Putting People First. Smart Cities and Communities. (2020). US Department of Transportation. https://its.dot.gov/smartcities/SmartCities.pdf (accessed 19.02.2024).

Qian, X. (2023). Evaluation on sustainable development of fire safety management policies in smart cities based on big data. *Mathematical Biosciences and Engineering*, 20(9), 17003–17017.

Qiu, J., Cao, J., Gu, X.,. Wang, Z., Liang, Z. (2023). Design of an evaluation system for disruptive technologies to benefit smart cities. *Sustainability*, 15(11), 9109.

Raj, A., Shetty, S. D. (2024). Smart parking systems technologies, tools, and challenges for implementing in a smart city environment: A survey based on IoT & ML perspective. *International Journal of Machine Learning and Cybernetics*, 15(7), 1–22.

Rui, L. (2017). Smart environment protection promotes development of smart cities. https://ggim.un.org/meetings/2017-Kunming/documents/Session%208%20-%20Liu%20Rui.pdf (accessed 19.02.2024).

Salleh, M. S. M., Fahmy-Abdullah, M., Sufahani, S. F., Bin Ali, M. K. (2022). Smart cities with smart environment. In Kaiser, M. S., Ray, K., Bandyopadhyay, A., Jacob, K., Long, K. S. (eds), *Proceedings of the third international conference on trends in computational and cognitive engineering*. Lecture Notes in Networks and Systems, vol. 348. Singapore: Springer. https://doi.org/10.1007/978-981-16-7597-3_22.

Samuel, R., Ponmani, S., Sophia, J. J., Mathew, D. M., Samuel, P. J. (2022). Smart living: Role of the internet of everything and the challenges. *Internet of Everything: Smart Sensing Technologies*, 1–30.

Santosa, I., Supangkat, S. H., Arman, A. A. (2023). Value of smart city services in improving the quality of life: A literature review. *10th International Conference on ICT for Smart Society*, 6–7 September 2023, Bandung, Indonesia.

Sharifi, A., Allam, Z., Bibri, S. E., Khavarian-Garmsir, A. R. (2024). Smart cities and sustainable development goals (SDGs): A systematic literature review of co-benefits and trade-offs. *Cities*, 146, 104659.

Shi, D., Cao, X. (2022). Research on the effectiveness of government governance in the context of smart cities. *Proceedings of SPIE: The International Society for Optical Engineering*, 12165, 121651F.

Sierdovski, M., Pilatti, L. A., Rubbo, P., Bittencourt, J. V. M., Pagani, R. N. (2022). Support from governance, leadership and smart people in the context of smart cities development. *ICE/ITMC 2022 and IAMOT 2022 Joint Conference*, 19–23 June 2022, Nancy, France.

Smart Cities and Inclusive Growth. (2020). Ministry of Land, Infrastructure and Transport, Korea. www.oecd.org/cfe/cities/OECD_Policy_Paper_Smart_Cities_and_Inclusive_Growth.pdf (accessed 19.02.2024).

Smart Governance Manual. (2018). European Union Regional Development Fund. https://programme2014-20.interreg-central.eu/Content.Node/UGB/HZI-Smart-Governance-Manual-Short-2019-ENG-WEB.PDF (accessed 19.02.2024).

Sobnath, D., Rehman, I. U., Nasralla, M. M. (2020). Smart cities to improve mobility and quality of life of the visually impaired. In *EAI/Springer Innovations in Communication and Computing*, Berlin: Springer, 3–28.

Sokołowski, M. M., Visvizi, A. (2023). Exploring the smart cities: Energy communities nexus – Issues and challenges. In *Routledge handbook of energy communities and smart cities*, 1–10.

Sovacool, B. K., Furszyfer Del Rio, D. D. (2020). Smart home technologies in Europe: A critical review of concepts, benefits, risks and policies. *Renewable and Sustainable Energy Reviews*, 120, 109663.

Šurdonja, S., Giuffrè, T., Deluka-Tibljaš, A. (2020). Smart mobility solutions-necessary precondition for a well-functioning smart city. *Transportation Research Procedia*, 45, 604–611.

Syalianda, S. I., Kusumastuti, R. D. (2021). Implementation of smart city concept: A case of Jakarta Smart City, Indonesia. *IOP Conference Series: Earth and Environmental Science*, 716(1), 012128.

Tyas, W. P., Nugroho, P., Sariffudin, S., Purba, N. G., Riswandha, Y., Sitorus, G. H. I. (2019). Applying smart economy of smart cities in developing world: Learnt from Indonesia's home based enterprises. *OP Conference Series: Earth and Environmental Science*, 248, 012078.

Vătăşoiu, R.-I., Brătulescu, R.-A., Mitroi, S.-A., Tudor, A.-M., Vintilă, A.-G. (2023). The importance of security and safety in a smart city. *Smart Innovation, Systems and Technologies*, 321, 11–23.

Vitálišová, K., Sýkorová, K., Koróny, S., Rojíková, D. (2022). Benefits and obstacles of smart governance in cities. *Lecture Notes of the Institute for Computer Sciences, Social-Informatics and Telecommunications Engineering*, 442, 366–380.

Wang, M., Zhou, T. (2023). Does smart city implementation improve the subjective quality of life? Evidence from China. *Technology in Society*, 72, 102161.

Wielicka-Gańczarczyk, K., Jonek-Kowalska, I. (2023). Involvement of local authorities in the protection of residents' health in the light of the smart city concept on the example of Polish cities. *Smart Cities*, 6(2), 744–763.

Wielicka-Gańczarczyk, K., Jonek-Kowalska, I. (2023). Perceptions and attitudes toward risks of city administration employees in the context of smart city management. *Smart Cities*, 6(3), 1325–1344.

Wilson, C., Hargreaves, T., Hauxwell-Baldwin, R. (2017). Benefits and risks of smart home technologies. *Energy Policy*, 103, 72–83.

Wirsbinna, A. (2021). The evaluation of economic benefits of smart city initiatives: A category approach. *Scientia International Economic Review*, 1(1), 3240.

Wolniak, R., Jonek-Kowalska, I. (2021). The level of the quality of life in the city and its monitoring. *Innovation: The European Journal of Social Science Research*, 34(3), 376–398.

Wolniak, R., Jonek-Kowalska, I. (2021). The quality of service to residents by public administration on the example of municipal offices in Poland. *Administratie si Management Public*, 37, 132–150.

Youssef, A., Hajek, P. (2022). The role of smart economy in developing smart cities. https://dk.upce.cz/bitstream/handle/10195/79089/Smart_Economy_paper_draft_-v2.pdf?sequence=1 (accessed 19.02.2024).

Zhong, Z., Zeng, Y., Zhao, X., Zhang, S. (2024). The social benefits resulting from electric vehicle smart charging balancing economy and decarbonization, *Transport Policy*, 147, 113–124.

3 Dimensions of exclusion in a Smart City

3.1 The essence and causes of exclusion in a Smart City

In Smart Cities, innovative processes aim to improve the quality of life of residents through the use of advanced technologies. However, despite the positive aspects, there are also some challenges associated with this dynamic development that may lead to potential exclusions. One of the main factors is unequal access to modern technological solutions. Residents in lower socio-economic status areas may face difficulties in using smart services due to lack of access to necessary infrastructure, such as high-speed internet or electronic equipment. This phenomenon may reinforce existing social inequalities by increasing the distance between different social groups (Kim, 2023).

Another important aspect is the issue of privacy. Implementing monitoring systems and collecting data to optimize the functioning of the city may raise concerns about the privacy of individuals. People, who fear excessive surveillance or inappropriate use of collected information, may choose not to use certain services, which affects their full participation in a technology-based society. Additionally, automation processes may lead to job losses, especially in sectors that traditionally employ lower-skilled workers. Without appropriate retraining and support programmes, people affected by job loss may find themselves on the margins of the digital society (Kim et al., 2018).

It is necessary to take these aspects into account and work closely with the community when planning and implementing technological innovations in Smart Cities. The priority should be to ensure equal access to the benefits of modern solutions and to minimize potential negative effects in order to build a more sustainable and inclusive society (Alidadi and Sharifi, 2022).

DOI: 10.4324/9781003499992-4

Despite the desire to create more effective and integrated systems, there are also certain reasons for exclusion in the area of smart governance, which may result in unequal access to the benefits of Smart City solutions (Table 3.1). One of the key factors is unequal access to technology. People with lower economic status or living in areas with weaker infrastructure may encounter difficulties in using modern services provided as part of smart governance. This phenomenon may lead to a digital social divide, where certain population groups are excluded from enjoying the benefits of Smart City solutions (Choi et al., 2021; van Gils and Bailey, 2023).

Another important cause of exclusion in the area of smart governance is the lack of appropriate digital skills among part of society. Even when technology is available, not all residents may be able to use it effectively. Older people, those with low professional qualifications or with limited digital education may be left out of modern solutions, which increases the likelihood of social exclusion.

Moreover, the aspect of data security constitutes a significant challenge for smart governance. Concerns about privacy and the risk of data misuse may encourage some groups of society to reject

Table 3.1 Reasons for exclusions in Smart Cities in the area of Smart Governance

Cause of exclusion	*Characteristics*
Low digital knowledge of residents	The community's lack of technology skills makes it difficult to participate in smart governance platforms and services.
Lack of access to local authority technology	Residents in areas with poor digital infrastructure may be excluded from using online services and tracking government activities.
Lack of participation of local communities	The lack of active participation of residents in decision-making processes leads to low representativeness and understanding of the community's needs.
Lack of transparency of administration activities	Lack of access to full information about the activities of local authorities may lead to community distrust in smart governance.
Insufficient digital security	Concerns about data security and privacy may discourage residents from using smart governance services and platforms.

Source: Authors' own work based on Kim (2023); Kim et al. (2018); Choi et al. (2021); Herrschel (2017); Van Gils and Bailey (2023); Correia et al. (2021).

participation in Smart City management systems. Lack of trust in the mechanisms of collecting, processing and storing data may lead to resistance towards modern solutions. Additionally, the lack of transparency in decision-making processes related to smart governance may lead to social exclusion. If decisions about the development of Smart Cities are made without adequately taking into account the opinions and needs of different social groups, there is a risk that some communities will be left behind, increasing social inequalities.

In the area of smart economy, in the process of striving to create sustainable urban economies, obstacles may result from problems related to exclusion and inequalities in access to modern solutions (Table 3.2). One of the fundamental problems is digital inequality in terms of access to technology. People and companies that do not have sufficient access to modern digital tools and platforms may be excluded from participating in smart economic initiatives. This phenomenon may deepen existing economic inequalities, leading to disparities between enterprises and economic units (Vinod Kumar et al., 2021).

The lack of technological infrastructure in some geographical areas is another significant cause of exclusion in the smart economy. Towns and their parts with poor telecommunications or energy infrastructure

Table 3.2 Reasons for exclusions in Smart Cities in the area of Smart Economy

Cause of exclusion	*Characteristics*
Digital inequalities in market access	Lack of equal access to technology may lead to the exclusion of certain social groups from the benefits of a smart economy.
Lack of support for local businesses	Smaller companies may be excluded from using technology, leading to increased economic inequality in Smart Cities.
Low digital competence of entrepreneurs	The lack of ability to use modern tools may limit the innovation and competitiveness of local companies.
Insufficient funding for digital initiatives	Lack of financial resources may limit the development of smart economy projects, especially those aimed at sustainable development.
Lack of integration between economic sectors	Lack of cooperation between sectors can lead to isolation of activities, which hinders the effective use of technology in the economy.

Source: Authors' own work based on Kim (2023); Kim et al. (2018); Vinord Kumar et al. (2021); Mboup et al. (2017); Wiig (2016); Makushin et al. (2016); Daila (2013).

may encounter difficulties in using advanced technologies, which hampers the development of the local economy based on smart solutions (Mboup et al., 2017).

Insufficient education and lack of digital skills among workers and entrepreneurs pose another challenge. The need to adapt to a rapidly changing economic environment based on technology may generate social exclusion for those, who are unable to adapt to the new requirements of the labour market. Additionally, data security and privacy issues may impact participation in smart economic solutions. Fear of data misuse or unauthorized access to business information may cause resistance towards technology, which, in turn, leads to social exclusion in the smart economy.

In the case of smart mobility (Table 3.3), one of the main exclusion factors is unequal access to modern means of transport. People from a lower economic status or living in areas with poor transport infrastructure may experience difficulties using smart mobility services. This phenomenon may lead to deepening existing social inequalities, where some residents are excluded from using modern means of transport. The lack of technological infrastructure in some urban

Table 3.3 Reasons for exclusions in Smart Cities in the area of Smart Mobility

Cause of exclusion	*Characteristics*
Insufficient transportation infrastructure	Lack of adequate infrastructure may limit the use of smart mobility solutions, especially for rural areas.
Low availability of public transport	The lack of effective public transport makes it difficult for residents to use modern mobility services in a Smart City.
Lack of access to mobile technology	People without access to smartphones or other mobile devices may be excluded from using many smart mobility services.
Low acceptance of autonomous technologies	Public lack of trust in autonomous vehicles may delay the adoption of innovations in the area of smart mobility.
Insufficient investment in infrastructure	Lack of appropriate investments may lead to neglecting the development of smart transport solutions in Smart Cities.

Source: Authors' own work based on Kim (2023); Kim et al. (2018); Bridgman et al. (2022); Rocha et al. (2021); Sourbatu and Behrendt (2021); Butler et al. (2021); Leviäkangas and Ahonen (2021); Bosch et al. (2021); Vinod Kumar et al. (2021); Zhang et al. (2020); Butler et al. (2020); Sagaris (2020); Growth (2019).

areas is another important reason for exclusions in the area of smart mobility. Towns with a poor road network, lack of safe bicycle paths or insufficient number of charging stations for electric vehicles may limit the availability of smart forms of transport (Rocvha et al., 2021; Bridgman et al., 2022).

Insufficient education of residents about modern mobility options and lack of knowledge of the benefits of smart transport solutions is another exclusion factor. People unaware of available options may remain distant from using smart mobility services, leading to further inequalities in access to modern modes of transport (Sourbati and Behrendt, 2021; Zhang et al., 2020; Sagaris, 2020).

It is also worth paying attention to security aspects, both physical and cybernetic, which are a significant challenge in the context of smart mobility. Fears of accidents, data theft or cyber-attacks on transport systems may cause reluctance towards modern solutions, especially among those groups of society that feel threatened or insecure.

One of the key factors of exclusion in the field of smart environment (Table 3.4) is unequal access to ecological innovations. Residents of

Table 3.4 Reasons for exclusions in Smart Cities in the area of Smart Environment

Cause of exclusion	*Characteristics*
Lack of ecological awareness of society	Low ecological awareness may lead to a lack of interest in using smart pro-ecological solutions.
Lack of environmental monitoring infrastructure	An insufficient number of sensors and monitors makes it difficult to effectively monitor the state of the environment in a Smart City.
Lack of motivation to participate in ecological projects	The lack of incentives to participate in pro-ecological projects may lead to low public involvement in the smart environment.
Conflicts of interest in natural resource management	Conflicts in the management of natural resources can lead to inconsistent ecological actions, which hinders the development of a smart environment.
Lack of access to green spaces	Lack of access to green areas and parks may lead to a lack of health benefits resulting from smart ecological solutions.

Source: Authors' own work based on Kim (2023); Kim et al. (2018); Mello Rose et al. (2022); Alidadi and Sharifi (2022); Praharaj (2021); Bisoyi et al. (2021); Jayashree et al. (2019); Sengupta and Sengupta (2022).

areas with a lower economic status may encounter difficulties in using smart environmental solutions, such as air quality monitoring systems or modern waste management methods. This phenomenon may contribute to widening environmental inequities when certain communities are deprived of the benefits of green innovations (Mello Rose et al., 2022).

The lack of ecological infrastructure in some urban areas is another important exclusion factor. An insufficient number of parks, a lack of green spaces or an insufficient number of places for recycling may limit the possibilities of using smart environmental solutions for residents of a given area. Insufficient ecological education of society is another factor hindering access to a smart environment. Lack of awareness about the benefits of environmental protection or ways to reduce one's impact on the planet may lead to ignoring modern environmental initiatives (Alidadi and Sharifi, 2022).

In addition, issues related to the costs of implementing smart environmental solutions may constitute a barrier for certain groups of society. People with lower incomes may be excluded from the use of advanced ecological technologies, which generates inequalities in access to environmental innovations.

One of the key factors of exclusion in the area of smart living (Table 3.5) is economic inequality, which may affect access to

Table 3.5 Reasons for exclusions in Smart Cities in the area of Smart Living

Cause of exclusion	*Characteristics*
Insufficient housing infrastructure	Lack of access to smart systems in buildings makes it difficult for residents to use smart home services.
Lack of access to digital education	Lack of ability to use modern technologies may limit the use of smart solutions in everyday life.
Lack of access to online health care	Lack of access to remote medical services may lead to inequality in access to health care in a Smart City.
Low acceptance of smart home systems	Lack of social acceptance for smart home systems may delay their adoption in smart living.
Insufficient safety of residents	The lack of appropriate security systems may limit residents' trust in smart solutions in the area of smart living.

Source: Authors' own work based on Kim (2023); Kim et al. (2018); Bordini et al. (2020); Napoles et al. (2021); Götzelmann and Kreimeier (2020); Bolay (2020); Parthasarathy and Sastry (2019); Schulze and Zirk (2014); Lamonaca and Batel (2023).

smart services related to everyday life. Lower-income residents may encounter difficulties in using modern solutions, such as smart housing or energy management systems, which increases inequalities in living standards (Napoles et al., 2021).

Lack of access to modern technologies in some urban areas is another important reason for exclusions in the area of smart living. Apartments or neighbourhoods with limited access to high-speed internet or telecommunications infrastructure may be excluded from using services that rely on advanced technologies (Parathasarathy and Sastry, 2019). Insufficient digital education and lack of technology skills pose another challenge. People, who are not familiar with using modern devices or applications, may be excluded from using smart solutions that improve everyday life, which leads to the digital divide (Bolay, 2020). Data security and privacy concerns are also important exclusion factors in the smart living space. People concerned about data abuse or privacy violations may reject the use of smart technologies at home, which generates inequalities in access to innovative solutions.

In the area of smart people, despite the aspiration to create a more integrated community and tailored to the needs of individuals, it is possible to identify reasons for exclusion that may lead to inequalities in access to Smart City solutions. One of the key factors of exclusion is inequality of access to digital education. People with a lower level of education or limited access to modern technologies may encounter difficulties in using smart social services, such as e-government or educational platforms. This phenomenon may contribute to deepening inequalities in access to information and development opportunities for various social groups (Zhou et al., Sidani et al., 2022).

Lack of equality in access to online health care is another important factor of exclusion (Table 3.6). People with limited internet access or insufficient medical education may be deprived of the benefits of smart health solutions, such as telemedicine or e-prescriptions. Insufficient public awareness of the benefits of modern social solutions is another factor hindering participation in the digital society. Lack of understanding of the opportunities offered by Smart Cities may lead to social exclusion and lack of participation in modern social initiatives (Malek et al., 2021; Alizadeh and Sharifi, 2023; Masucci et al., 2020).

Table 3.6 Reasons for exclusions in Smart Cities in the area of Smart People

Cause of exclusion	*Characteristics*
Low digital competence of society	Lack of technology skills makes it difficult for society to use smart solutions in various areas of Smart Cities.
Lack of access to digital education	Lack of access to digital education limits the development of skills necessary to use technology in a Smart City.
Low awareness of the benefits of technology	Lack of understanding of the benefits of smart solutions may discourage society from accepting and using them.
Lack of participation in participatory processes	The lack of active public participation in decision-making and participatory processes leads to low involvement in the development of Smart Cities.
Lack of access to basic digital services	Lack of access to basic online services can lead to social isolation, especially in rural areas.

Source: Authors' own work based on Kim (2023); Kim et al. (2018); Zhou et al. (2023); Sidani et al. (2022); Malek et al. (2021); Sarkar (2019); McKenna (2017); Alizadeh and Sharifi (2023); Masucci et al. (2020); Sagaris (2020).

3.2 Research methodology of exclusion in a Smart City

Exclusions in Smart Cities are a complex phenomenon covering various aspects of social life. This concept refers to a situation, in which certain groups of city residents do not have equal access to the benefits resulting from the dynamic development of technology and smart urban solutions. In the context of Smart Cities, where innovative technologies are a key element in improving the quality of life, exclusions can take various forms. Table 3.7 presents the most important types of exclusions occurring in Smart Cities.

Exclusions in Smart Cities have a negative impact on the quality of life of residents and deepen social inequalities. Therefore, it is important for city authorities and other entities responsible for the development of Smart Cities to take action to counteract these exclusions.

Digital exclusion is a complex phenomenon that includes lack of access to the internet, digital devices or the skills to use them. This is not only a matter of physical access to infrastructure, but also a

Table 3.7 Characteristics of the main types of exclusions in a Smart City and ways to counteract these exclusions

Exclusion type	*Characteristics*	*Counteracting methods*
Digital exclusion	Lack of access to the internet, digital devices or the ability to use them.	• Providing free WiFi access in public spaces. • Providing low-cost or free digital devices. • Organizing training in computer and internet use.
Economic exclusion	Inability to use Smart City services due to high costs.	• Introduction of subsidies for people with low incomes. • Development of discount programmes for specific social groups. • Providing free or low-cost alternatives to paid services.
Social exclusion	Inability to participate in the life of the Smart City community due to marginalization or isolation.	• Organizing integration events for various social groups. • Supporting the activities of non-governmental organizations dealing with social welfare. • Creating online platforms enabling communication and exchange of information.
Educational exclusion	Lack of access to education and training necessary to use Smart City technology.	• Providing free access to online education. • Organizing courses and training in the field of digital skills. • Supporting educational programmes for children and youth.
Linguistic exclusion	Lack of knowledge of the language, in which Smart City services are available.	• Translating user interfaces into various languages. • Providing information materials in different languages. • Providing translators for people, who do not speak the dominant language.
Spatial exclusion	Lack of access to Smart City infrastructure, e.g. due to location or disability.	• Taking into account the needs of people with disabilities in infrastructure design. • Ensuring the availability of public transport. • Developing digital infrastructure in less accessible areas.

Table 3.7 (Continued)

Exclusion type	*Characteristics*	*Counteracting methods*
Political exclusion	Inability to participate in decision-making processes regarding the development of a Smart City.	• Increasing the transparency of decision-making processes. • Ensuring residents' participation in public consultations. • Creating platforms for dialog between authorities and residents.

Source: Authors' own work based on Kim (2023); Kim et al. (2018); Napoles et al. (2021); Kim (2022); Bleja et al. (2020); Valdez et al. (2020); László and Ráhel (2016); Chib et al. (2022); Torres-Gonzalez et al. (2023); Allam and Newman (2023); Bunyan and Collins (2013); Liu et al. (2018); Amaro Agudo and Navarro Mateu (2013); Park (2024); Zhou et al. (2022); Owiedo (2021); Zembri (2021); Mbugabo and Orock (2012); Kundu (2020); Cheema (2020); Malek et al. (2021); Alidadi and Sharifi (2022); Napoles et al. (2021).

problem related to inequalities in digital knowledge and skills. Digital exclusion leads to social marginalization and becomes an important factor hindering the use of many services offered by Smart Cities.

Lack of access to the internet limits the ability to participate in the information society, where communication, remote work, online education and access to information are becoming a standard. People, who do not have access to broadband internet, are at a disadvantage and do not have equal opportunities to use modern solutions. Additionally, the lack of appropriate digital devices, such as smartphones, tablets or computers, prevents the use of numerous services available within a Smart City. Mobile applications, e-service platforms and monitoring systems require equipment, which means that people deprived of these resources are excluded from full participation in social life (Kim et al., 2018).

The adverse effects of digital exclusion also include the lack of skills in using modern technologies. In a society based on digital solutions, the ability to use a computer, use applications and move safely in virtual space becomes a key competence. People without these skills encounter difficulties not only in using Smart City services, but also in everyday functioning. In order to effectively counteract digital exclusion in the context of a Smart City, it is necessary to develop educational programmes, invest in technological infrastructure and promote equal access to the internet. Only by eliminating barriers related to lack of access to technology, digital education and promoting digital inclusion can Smart Cities be created, in which the benefits of technological progress are available to all residents.

Economic exclusion is a significant challenge in the context of a Smart City, which assumes development based on advanced technologies. This phenomenon involves the inability to use Smart City services due to high costs, which primarily affects low-income people, who cannot afford to buy digital devices or pay subscriptions for digital services (Vinod Kumar et al., 2021).

In Smart Cities, where innovative technologies are an integral part of everyday life, low-income people may be excluded from full participation in digital society. The high costs of purchasing smartphones, tablets or computers and paying subscriptions for broadband internet may constitute a significant financial burden for these social groups. This leads to a situation, in which people with lower incomes not only do not have access to modern Smart City solutions, but are also deprived of the benefits resulting from them. They cannot use applications that facilitate everyday functioning, they do not participate in remote education or work, which contributes to increasing social inequalities (Napoles et al., 2021).

To effectively counteract economic exclusion in a Smart City, it is necessary to introduce policies aimed at reducing the costs of access to technology. This could include subsidies for the purchase of digital devices, financial assistance programmes to help pay for digital services, or investment in infrastructure to enable free access to broadband internet in public places. Only by eliminating financial barriers related to participation in the digital society can a Smart City be created that is open and accessible to all residents, regardless of their economic situation.

Social exclusion in the context of a Smart City refers to a situation, in which certain groups of people experience an inability to participate in the life of a Smart City community due to their marginalization or isolation. This phenomenon can be caused by various factors, such as disability, poverty, ethnicity or language. People with disabilities may encounter challenges in accessing Smart City infrastructure, which does not always take special needs into account. The lack of appropriate facilities, such as accessibility for people in wheelchairs or communication systems adapted to various types of disabilities, may lead to the isolation of these people in a smart environment (Malek et al., 2021; Alizadeh and Sharifi, 2023; Wolniak and Skotnicka-Zasadzień, 2019).

Poverty can also be a cause of social exclusion in a Smart City, as people with low incomes may have limited access to modern technologies, digital education or even basic city services. Lack of access to the internet or lack of funds to purchase digital devices leads to the loss of the ability to use the services offered by Smart Cities. Ethnic or cultural origin and language are other factors influencing social exclusion in a Smart City. If infrastructure and services are not adapted to the diversity of society, groups with specific ethnic or linguistic needs may be omitted, leading to their social isolation (Masucci et al., 2020; Sidani et al., 2022).

In order to effectively counteract social exclusion in a Smart City, it is necessary to take into account social and cultural diversity in planning the city's development. Implementing inclusive technological solutions tailored to different social groups and promoting equal access to city services are key elements of creating Smart Cities, in which every resident has the opportunity to actively participate in the community.

Educational exclusion in the context of a Smart City is a situation, in which certain social groups experience a lack of access to education and training necessary to fully use Smart City technologies. This phenomenon may lead to limited opportunities for personal and professional development and may also deepen existing social inequalities.

People, who do not have access to appropriate digital education, face difficulties in understanding and using advanced technologies that constitute the foundation of a Smart City. Lack of basic knowledge on how to use digital devices, use applications or move safely in virtual space may constitute a barrier preventing full participation in the life of a city based on smart solutions. Limited access to education in the field of Smart City technologies affects not only digital skills, but also the development of professional competences. The modern labour market increasingly requires knowledge of modern tools and technologies, and the lack of adequate education may result in difficulties in obtaining a job, which, in turn, contributes to maintaining social inequalities (Liu et al., 2018; Amaro Agudo and Navarro Mateu, 2013).

Moreover, educational exclusion may exacerbate intergenerational differences as older generations encounter difficulties in adapting to a rapidly changing digital environment. People, who do not participate in the process of continuous digital learning, are at risk of exclusion, both socially and professionally. To counteract educational exclusion in a Smart City, it is necessary to create educational programmes that are adapted to various social groups. Investments in digital training, educational programmes for older people and financial support for low-income families are key to creating Smart Cities that are equally accessible to all residents.

Linguistic exclusion in the context of a Smart City means a situation, in which certain social groups experience difficulties due to a lack of knowledge of the language, in which Smart City services are available. This applies primarily to immigrants and people belonging to linguistic minorities, which may lead to limiting the benefits of innovative solutions offered by a Smart City. People, who do not speak the dominant language in a given community, may encounter difficulties in understanding information, using applications, or communicating with Smart City systems. Limited access to information in a language understandable to a given social group leads to the exclusion of these people from full participation in the life of a city based on advanced technologies.

Immigrants and people, who belong to linguistic minorities, may encounter difficulties in accessing public, educational and health care services, because communication and information availability are insufficiently adapted to the linguistic diversity of society. This phenomenon can lead to social isolation, limit integration opportunities and make it difficult to use smart solutions. To effectively counteract linguistic exclusion in a Smart City, it is important to take multilingualism into account in the design of services and applications. Translations, multilingual interfaces and adapting messages to different language groups are key to ensuring equal access to the benefits of Smart City technology. Moreover, promoting linguistic inclusion can help to increase understanding and cooperation between different communities in Smart Cities.

Spatial exclusion in the context of a Smart City is a situation, in which certain social groups experience a lack of access to Smart City infrastructure due to geographical location or disability. This phenomenon may lead to the loss of equal access to advanced technologies and services offered by Smart Cities (Kolotouchkina et al., 2022). Geographic location may be an important factor determining access to innovative Smart City solutions. Residents of rural areas, remote suburbs or degraded urban neighbourhoods may encounter difficulties in using Smart City infrastructure, which is often concentrated in central city areas. Lack of access to modern means of transport, smart energy systems or even access to the internet in areas with poor infrastructure can lead to social and economic exclusion (Park, 2024; Oviedo, 2021).

Physical or sensory disabilities are another exclusion factor when the Smart City infrastructure is not adapted to special needs. The lack of facilities, such as elevators for wheelchair users, communication systems for the blind or the availability of public spaces for people with disabilities, may prevent the full use of modern Smart City solutions (Wolniak and Skotnicka-Zasadzień, 2021; Smith, 2021).

To counteract spatial exclusion in a Smart City, it is necessary to distribute infrastructure and services evenly in different parts of the city and to take into account the needs of people with various types of disabilities in the urban planning process (Svelec et al., 2020; Wolniak and Skotnicka-Zasadzień, 2018). Investments in infrastructure development in marginalized areas and providing access to advanced technologies in less populated areas are key to creating Smart Cities that are accessible to all residents, regardless of their location or physical abilities.

Political exclusion in the context of a Smart City is a situation, in which certain social groups experience the inability to participate in decision-making processes regarding the development of Smart Cities. This phenomenon may result from various factors, such as lack of representation, limited access to information or discriminatory practices, which results in the exclusion of these social groups from shaping strategies and policies related to a Smart City. Individuals or groups experiencing political exclusion may encounter difficulties in accessing city decision-making structures, where key decisions regarding the development of Smart City infrastructure and technologies are made. Lack of representation in decision-making bodies or limited access to information on planned investments may lead to their needs and prospects being omitted in planning processes (Mbugabo and Orock, 2012; Kundu, 2020).

Additionally, political exclusion may result from unequal access to information resources, which limits the ability of individuals or communities to consciously participate in public debates or consultations regarding Smart Cities. The lack of equal access to information may contribute to the development of information asymmetries, which hinders effective participation in decision-making processes. To counteract political exclusion in a Smart City, it is necessary to introduce transparent and inclusive social participation mechanisms. This includes broader representation of various social groups in decision-making bodies, access to clear information on Smart City projects, as well as organizing public consultations and dialog that take into account the diverse perspectives and needs of residents. In this way, Smart Cities can be created, in which decisions are made in cooperation with the entire society, thus eliminating political exclusion (Kim, 2022; Cheema, 2020).

In order to effectively counteract exclusions in Smart Cities, a holistic approach is necessary, taking into account economic, social, digital and ecological aspects. The introduction of equality and education policies, as well as investments in technology access infrastructure, can contribute to the creation of Smart Cities, in which the benefits of technological progress are available to all residents.

Analysis of social exclusion problems in Smart Cities is an important issue that requires the attention of various entities, including public institutions, enterprises, researchers and local communities. This approach is key to creating Smart Cities that are more equitable, sustainable and accessible for all residents. Some of the main entities important in the analysis of exclusion problems in Smart Cities are public institutions, such as local governments and the government. They are responsible for shaping urban policies, implementing innovative solutions and ensuring equal access to urban resources. This analysis allows you to identify areas, where there is social inequality, lack of access to basic services or infrastructure, which allows you to take targeted actions.

Enterprises should also engage in exclusion analysis in Smart Cities, for both ethical and business reasons. Creating products and services that take into account social diversity contributes not only to increasing market reach but also to building a positive image of the company. This analysis allows companies to adapt their strategies to the real needs of the community, eliminating the barrier of access to modern technologies.

Researchers play a key role in analysing exclusion problems in Smart Cities by conducting scientific research, analysing data and identifying social trends. It is because of their works (Kim, 2023; Kim et al., 2018; Alidadi et al., 2022; Napoles et al., 2021; Malek et al., 2021; Alizadeh and Sharifi, 2023) that we can understand the deeper causes of exclusions and propose innovative solutions. Cooperation between researchers, public institutions and enterprises allows for the effective implementation of research results in practice.

Local communities also take part in the analysis of exclusion problems, often through citizen initiatives and participation in participatory processes. Local knowledge and public engagement are essential to identifying area-specific needs and effectively solving problems at the micro-community level.

How the analysis of exclusion problems in Smart Cities is carried out depends on many factors, including available technologies, data and the specificity of a given community. The analysis may include collecting demographic data, analysing the availability of services, monitoring inequalities in access to technology, or assessing the effects of introduced innovations on various social groups.

Analysing the problems of exclusion in Smart Cities requires a multi-sectoral approach, cooperation between different actors and taking into account the perspective of local communities. It is, therefore, possible to create Smart Cities that not only take advantage of modern technologies but also ensure that innovations benefit all residents.

Despite significant progress toward the development of Smart Cities, there are many areas related to social exclusion that are not sufficiently studied and described in the literature. One such insufficiently explored thread is the problem of access to modern technology by the elderly. As Smart Cities become increasingly dependent on digital technologies, the elderly can often be overlooked or on the margins of innovative solutions, leading to further digital segregation.

Another aspect that requires more detailed analysis is the impact of Smart Cities on local labour markets. Despite promises of creating new jobs in the technology sector, it is not always clear what the impact will be on traditional economic sectors. Research on which social groups may lose out due to the digital transformation is essential to avoid further increases in social inequality.

Another aspect that deserves more attention is the privacy and data security issues in Smart Cities. Technologically developed cities are collecting huge amounts of data on their residents in the context of digital services. Research into how this data is stored, processed and secured is critical to protecting citizens' privacy and minimizing the risk of abuse.

In addition, aspects related to citizen participation in Smart City decision-making processes require a more comprehensive approach. It is worth considering whether there are differences in access to public participation based on socio-economic status, education or age. Open and inclusive public participation is crucial to creating Smart Cities that reflect the diversity of their residents' needs.

Bibliography

Acevedo, E. A. C., Duque, A. J. (2021). Guidelines to define a regulatory proposal in the transition and inclusion of non-conventional renewable energies in Colombia and its role in the development of smart cities. *Smart Innovation, Systems and Technologies*, 233, 227–237.

Alidadi, M., Sharifi, A. (2022). The extent of inclusion of smart city indicators in existing urban sustainability assessment tools. *Urban Climate Adaptation and Mitigation*, 175–198.

Alizadeh, H., Sharifi, A. (2023). Toward a societal smart city: Clarifying the social justice dimension of smart cities. *Sustainable Cities and Society*, 95, 104612.

Allam, Z., Newman, P. (2023). Smart cultural and inclusive cities: How smart city can help urban culture and inclusion. *Cities and Nature*, F345, 77–99.

Amaro Agudo, A., Navarro Mateu, D. (2013). External agents, educational cities and inclusion: Axes of change in the classrooms. *Historia y Comunicacion Social*, 18, 449–460.

Bisoyi, B., Nayak, B., Das, B., Pasumarti, S. S. (2021). Urban resilience and inclusion of smart cities in the transformation process for sustainable development: Critical deflections on the smart city of Bhubaneswar in India. *Lecture Notes in Electrical Engineering*, 690, 149–160.

Bleja, J., Langer, H., Grossmann, U., Morz, E. (2020). Smart cities for everyone: Age and gender as potential exclusion factors. In *2020 IEEE European Technology and Engineering Management Summit*, 5–7 March 2020, Dortmund, Germany, 9111741.

Bolay, J.-C. (2020). When inclusion means smart city: Urban planning against poverty. *Advances in Intelligent Systems and Computing*, 1069, 283–299.

Bordini, R. H., Mascardi, V., Costantini, S., … Lespérance, Y., Ricci, A. (2020). Transcultural health-aware guides for the elderly. *CEUR Workshop Proceedings*, 2706, 135–146.

Borghetti, F., Longo, M., Mazzoncini, R., Cesarini, L., Contestabile, L. (2021). Relationship between railway stations and the territory: Case study in Lombardy – Italy for 15-min station. *International Journal of Transport Development and Integration*, 5(4), 367–378.

Bosch, E. R., Wybraniec, B., Lazzarini, B., Junyent, M. V., Crespo, À. G. (2021). The DIGNITY project: Toward a system of inclusive digital mobility in the Barcelona metropolitan area. *Transportation Research Procedia*, 58, 134–141.

Bridgman, J., Woodcock, A., Gut, K. (2022). How can gender smart mobility become a more intersectional form of mobility justice. In *Proceedings of the international conference on gender research*, 28-29 April 2022, Aveiro, Portugal. April, 65–71.

Bunyan, S., Collins, A. (2013). Digital exclusion despite digital accessibility: Empirical evidence from an English city. *Tijdschrift voor Economische en Sociale Geografie*, 104(5), 588–603.

Butler, L., Yigitcanlar, T., Paz, A. (2020). How can smart mobility innovations alleviate transportation disadvantage? Assembling a conceptual framework through a systematic review. *Applied Sciences*, 10(18), 6306.

Butler, L., Yigitcanlar, T., Paz, A. (2020). Smart urban mobility innovations: A comprehensive review and evaluation. *IEEE Access*, 8, 196034–196049.

Butler, L., Yigitcanlar, T., Paz, A. (2021). Barriers and risks of mobility-as-a-service (MaaS) adoption in cities: A systematic review of the literature. *Cities*, 109, 103036.

Cheema, S. (2020). Governance for urban services: Towards political and social inclusion in cities. *Advances in 21st Century Human Settlements*, 1–30

Chib, A., Alvarez, K., Todorovic, T. (2022). Critical perspectives on the smart city: Efficiency objectives vs inclusion ideals. *Journal of Urban Technology*, 29(4), 83–99.

Choi, J., Lee, S., Jamal, T. (2021). Smart Korea: Governance for smart justice during a global pandemic. *Journal of Sustainable Tourism*, 29(2–3), 540–549.

Correia, D., Feio, J., Teixeira, L., Lourenço Marques, J. (2021). The inclusion of citizens in smart cities policymaking: The potential role of development studies' participatory methodologies. *Lecture Notes in Computer Science*, 12782, 29–40.

Dalia, D. (2013). Jùjū, an enabling solution to guarantee digital inclusion in the perspective of the smart city. In *Proceedings of the international conference on engineering design*, 19–22 August 2013, Sungkyunkwan University, Seoul, DS75-04, 457–466.

Götzelmann, T., Kreimeier, J. (2020). Towards the inclusion of wheelchair users in smart city planning through virtual reality simulation. *ACM International Conference Proceeding Series*, 440–446, 3398008.

Groth, S. (2019). Multimodal divide: Reproduction of transport poverty in smart mobility trends. *Transportation Research Part A: Policy and Practice*, 125, 56–71.

Herrschel, T. (2017). Metropolitan elitism, marginalisation and the need for smartness in governing the regionalised city. *Territorio*, 83, 32–36.

Jayashree, P., Hamza, F., El Barachi, M., Gholami, G. (2019). Inclusion as an enabler to sustainable innovations in smart cities: A multi-level framework. In *4th International Conference on Smart and Sustainable Technologies*, 18–21 June, Bol (island of Brac) and Split (Croatia), 8783013.

Kim, D., Kwon, H.-Y., Jun, D., Hagen, L., Chun, S. A. (2018). Opportunities and challenges in the intelligent society: Smart cities, digital inclusion, and cybersecurity. *ACM International Conference Proceeding Series*, a131.

Kim, K. (2022). Exclusion and cooperation of the urban poor outside the institutional framework of the smart city: A case of Seoul. *Sustainability*, 14(20), 13159.

Kim, K. (2023). Inclusion, exclusion, and participation in digital polis: Double-edged development of poor urban communities in alternative smart city-making. *Advances in 21st Century Human Settlements*, 155–178.

Kolotouchkina, O., Barroso, C. L., Sánchez, J. L. M. (2022). Smart cities, the digital divide, and people with disabilities. *Cities*, 123, 103613.

Kundu, D. (2020). Political and social inclusion and local democracy in Indian cities: Case studies of Delhi and Bengaluru. *Advances in 21st Century Human Settlements*, 185–208.

Lamonaca, L., Batel, S. (2023). What of recognition justice? An empirical analysis of the role of recognition justice in social housing smart city projects in the Global North. In Living with *e*nergy *p*overty*:* Perspectives from the Global North and *South*, London: Routledge, 244–254.

László, G., Ráhel, C. (2016). Do smart cities intensify social exclusion? *Informacios Tarsadalom*, 16(3), 83–100.

Leviäkangas, P., Ahonen, V. (2021). The evolution of smart and intelligent mobility: A semantic and conceptual analysis. *International Journal of Technology*, 12(5), 1019–1029.

Liu, T., Holmes, K., Zhang, M. (2018). Better educational inclusion of migrant children in urban schools? Exploring the influences of the population control policy in large Chinese cities. *Asian Social Work and Policy Review*, 12(1), 54–62.

Makushkin, S. A., Kirillov, A. V., Novikov, V. S., Shaizhanov, M. K., Seidina, M. Z. (2016). Role of inclusion 'smart city' concept as a factor in improving the socio-economic performance of the territory. *International Journal of Economics and Financial Issues*, 6(1S), 152–156.

Malek, J. A., Lim, S. B., Yigitcanlar, T. (2021). Social inclusion indicators for building citizen-centric smart cities: A systematic literature review. *Sustainability*, 13(1), 1–29, 376.

Masucci, M., Pearsall, H., Wiig, A. (2020). The smart city conundrum for social justice: Youth perspectives on digital technologies and urban transformations. *Annals of the American Association of Geographers*, 110(2), 476–484.

Mboup, G., Diongue, M., Ndiaye, S. (2017). Smart city foundation: Driver of smart cities. *Advances in 21st Century Human Settlements*, 841–869.

Mbuagbo, O. T., Orock, R. T. (2012). Urban governance policies in Cameroon and the crisis of political exclusion: A case study of the city of Kumba. *Journal of Asian and African Studies*, 47(2), 176–189.

McKenna, H. P. (2017). Re-conceptualizing social inclusion in the context of 21st-century smart cities. *Social Inclusion and Usability of ICT-Enabled Services*, 49–66.

Mello Rose, F., Thiel, J., Grabher, G. (2022). Selective inclusion: Civil society involvement in the smart city ecology of Amsterdam. *European Urban and Regional Studies*, 29(3), 369–382.

Nápoles, V. M. P., Páez, D. G., Penelas, J. L. E., de Pablos, F. M., Gil, R. M. (2021). Social inclusion in smart cities. In *Handbook of smart cities*, Berlin: Springer, 469–514.

Oviedo, D. (2021). Making the links between accessibility, social and spatial inequality, and social exclusion: A framework for cities in Latin America. *Advances in Transport Policy and Planning*, 8, 135–172.

Park, I. K. (2024). Unravelling the relationship between spatial and social inclusion: Evidence from Korean cities and regions. *Local Environment*, 29(1), 21–39.

Parthasarathy, B., Sastry, B. (2019). Intelligence for place-making and social inclusion: Critiques and alternatives to India's smart cities mission. In *The new companion to urban design*, London: Routledge, 571–581.

Praharaj, S. (2021). Area-based urban renewal approach for smart cities development in India: Challenges of inclusion and sustainability. *Urban Planning*, 6(4), 202–215.

Rocha, N. P., Bastardo, R., Pavão, J., Queirós, A., Dias, A. (2021). Smart cities' applications to facilitate the mobility of older adults: A systematic review of the literature. *Applied Sciences*, 11(14), 6395.

Sagaris, L. (2020). Gendering smart mobilities in Latin America: Are 'smart cities' smart enough to improve social justice? In *Gendering smart mobilities*, London: Routledge, 229–250.

Sarkar, S. (2019). A mission to converge for inclusion? The smart city and the women of Seelampur. Media, Culture and Society, 41(3), 278–293.

Schulze, E., Zirk, A. (2014). Personalized smart environments to increase inclusion of people with Down's syndrome: Results of the requirement analysis. *Lecture Notes in Computer Science*, 8548, 144–147.

Sengupta, U. (2022). SDG-11 and smart cities: Contradictions and overlaps between social and environmental justice research agendas. *Frontiers in Sociology*, 7, 995603.

Sidani, D., Veglianti, E., Maroufkhani, P. (2022). Smart cities for a sustainable social inclusion strategy: A comparative study between Italy and Malaysia. *Pacific Asia Journal of the Association for Information Systems*, 14(2), 25–41.

Smith, R. O. (2021). Enabling smart cities to become even smarter through design for disability. *PerCom Workshops 2021*, 141, 9430949.

Sourbati, M., Behrendt, F. (2021). Smart mobility, age and data justice. *New Media and Society*, 23(6), 1398–1414.

Svelec, D., Bjelcic, N., Blazekovic, M. (2020). Smart cities as an opportunity and challenge for people with disabilities. In *43rd International Convention on Information, Communication and Electronic Technology*, 28 September–2 October, Opatija, Croatia, 456–461.

Torres-González, M. A., Hernández-Veleros, Z. S., Ramírez-Rosas, J. G., Guzmán-Escorza, L. E. (2023). Older adults in Mexico: Their inclusion in ICTs and smart cities. In *Management, t*echnology, and *e*conomic *g*rowth in *s*mart and *s*ustainable *c*ities, Hershey, PA: IGI Global, 111–132.

Valdez, A.-M., Wigley, E., Zanetti, O., Gillian, G. (2020). Learning lessons for avoiding the inadvertent exclusion of communities from smart city projects. In *Shaping smart for better cities: Rethinking and shaping relationships between urban space and digital technologies*, Cambridge, MA: Academic Press, 221–237.

van Gils, B. A. M., Bailey, A. (2023). Revisiting inclusion in smart cities: Infrastructural hybridization and the institutionalization of citizen participation in Bengaluru's peripheries. *International Journal of Urban Sciences*, 27(S1), 29–49.

Vinod Kumar, T. M., Sruthi Krishnan, V., Deepak Lawrence, K., Mohammed Firoz, C., Cyriac, S. (2021). Design of smart global economic community in Kattangal. *Advances in 21st Century Human Settlements*, 337–420.

Wiig, A. (2016). The empty rhetoric of the smart city: From digital inclusion to economic promotion in Philadelphia. *Urban Geography*, 37(4), 535–553.

Wolniak, R., Skotnicka-Zasadzień, B. (2018). Developing a model of factors influencing the quality of service for disabled customers in the conditions of sustainable development, illustrated by an example of the Silesian Voivodeship public administration. *Sustainability*, 10(7), 1–17.

Wolniak, R., Skotnicka-Zasadzień, B. (2021). Improvement of services for people with disabilities by public administration in Silesian Province Poland. *Sustainability*, 13(2), 1–16.

Wolniak, R., Skotnicka-Zasadzień, B., Zasadzień, M. (2019). Problems of the functioning of e-administration in the Silesian region of Poland from the perspective of a person with disabilities. *Transylvanian Review of Administrative S*ciences, 57E, 137–155,

Zembri, P. (2021). The challenges of accessibility in the city: Between spatial equity and social exclusion. *Territoire en Mouvement*, 47, 14–17.

Zhang, M., Zhao, P., Qiao, S. (2020). Smartness-induced transport inequality: Privacy concern, lacking knowledge of smartphone use and unequal access to transport information. *Transport Policy*, 99, 175–185.

Zhou, C., Zhan, M., An, X., Huang, X. (2022). Social inclusion concerning migrants in Guangzhou City and the spatial differentiation. *Sustainability*, 14(23), 15548.

Zhou, R. J., Chen, S., Zhang, B. (2023). Smart city construction and new-type urbanization quality improvement. *Scientific Reports*, 13(1), 21074.

4 The issue of exclusion in Smart City rankings

4.1 Characteristics of Smart City rankings

With the development of the Smart City concept and the implementation of Smart City solutions, attempts have begun to evaluate and rank cities aspiring to be smart. In theory and practice, a variety of rankings have emerged (Dashkevych and Portnov, 2023; Mokarrari and Torabi, 2022), in which cities were evaluated in specific categories, creating a hierarchy of international leaders in the creation and implementation of smart urban solutions (, 2023; Caird and Hallett, 2019; Caird, 2018).

Smart City evaluation, like the Smart City concept, has its supporters and opponents. Among the advantages of Smart City rankings is certainly the promotion of best practices to improve the quality of life of residents and the identification of cities (He, 2023) that are friendly to all urban stakeholders (Wang, 2023) and continuously strive for sustainable development (Mora et al., 2021).

The disadvantage of rankings and assessments, on the other hand, is the subjectivity of the evaluation, as they are done on the basis of expert criteria and scores (Mora et al., 2019). Furthermore, their results indicate the promotion of mainly large cities located in developed economies. This approach makes 'the strong even stronger' and 'the weak relatively weaker'. The selectivity of assessment areas in rankings can also result in the neglect of less attractive aspects of urban development and excessive focus on those aspects that are rated.

According to the authors of the monograph, however, Smart City rankings and assessments have more advantages than disadvantages, as they promote the noble cause of urban development and the important goal of improving life in modern cities (Banerjee et al.,

DOI: 10.4324/9781003499992-5

2023). They create room for competition (Ang-Tan and Ang, 2022). They disseminate good practices. They contribute to the promotion of innovation in the open access system (Parjanen and Rantala, 2021). Thus, they also unify the urban community around the shared values exhibited and assessed within each Smart City activity area.

Given that issues of exclusion are a decidedly less attractive aspect of Smart Cities (Lee et al., 2023; Caragliu and Del Bo, 2022), this monograph seeks to answer the questions: whether and how they are being highlighted in international Smart City assessments. To this end, this chapter describes 3 of the most popular Smart City assessment methods. Then, based on the review, an attempt was made to identify the aspects of exclusion that are taken into account in them. The following were selected for analysis:

- IESE Cities in Motion Index;
- Smart City Index;
- ISO 37120 Sustainable development of communities – Indicators for city services and quality of life.

Thus, the IESE Cities in Motion Index (Berrone and Ricart, 2022) is an international ranking developed by a team of specialists from the IESE Business School, one of the world's top business schools, part of Spain's University of Navarra. The ranking was created in 2014 and included 50 indicators to assess Smart Cities. In the following years, it was systematically developed. Currently, the evaluation in this ranking is based on 101 indicators. The evolution of the described evaluation shows the maturity of its creators and listening to the needs of the urban environment and stakeholders. It also illustrates the systematic pursuit of improvements in Smart City development, which could be an argument for the existence and use of such evaluation systems.

The IESE Cities in Motion Index includes a total of 9 evaluation areas. The indicators assigned to each area are shown in Tables 4.1–4.9.

According to the indicators in Table 4.1, the human capital framework primarily assesses the level of education, and this is done both in terms of school enrolment and education spending. Equally important are aspects of cultural life, such as museums, theatres and galleries. Thus, the factors assessed are both basic and elite in nature, related to the satisfaction of higher-order needs.

Table 4.1 Dimensions and indicators included in the IESE Cities in Motion Index (Human Capital)

Dimension	*Indicators*
Human capital	Secondary and higher education: proportion of population with secondary and higher education.
	Schools: number of public and private schools in a city.
	Business schools: number of business schools in the city included in the Financial Times TOP 100.
	Expenditure on education: annual private expenditure on education per capita.
	Expenditure on leisure and recreation: consumer expenditure on leisure and recreation as a percentage of GDP.
	Expenditure on leisure and recreation per capita: annual consumer expenditure on leisure and recreation per capita.
	Student mobility: international flow of mobile students at the tertiary level; number of students.
	Museums and art galleries: number of museums and art galleries in a city.
	Number of universities: number of TOP 500 universities.
	Theatres: number of theatres in a city.

Source: www.iese.edu/media/research/pdfs/ST-0633-E.pdf.

Nevertheless, the authors of the ranking devote a lot of space to social cohesion and issues directly related to exclusion (Table 4.2).

Nevertheless, the authors of the ranking devote a lot of space to social cohesion and issues directly related to exclusion (Table 4.2). In line with the above, indicators relating to exclusion based on gender, sexual orientation or racial affiliation can be found in the ranking. Moreover, it also includes elements related to slavery, violence and crime. This means that the creators of the compilation do not abstract from the dark sides of urban life. They additionally combine them with an assessment of the level of health care, wealth and happiness. All indicators determining Social Cohesion have a direct impact on the quality of urban life, as they are felt and interpreted by the residents themselves.

In the area concerning urban economics (Table 4.3), there are classic measures of society's wealth, based on the level of gross domestic product (GDP) and wage levels. You can also find indicators relating to the facilitators in doing business offered to entrepreneurs

Table 4.2 Dimensions and indicators included in the IESE Cities in Motion Index (Social Cohesion).

Dimension	*Indicators*
Social cohesion	Female-friendly: this variable indicates whether a city provides a friendly environment for women (on a scale of 1 to 5). Cities with a value of 1 have a more hostile environment for women; those with a value of 5 are very female-friendly. Hospitals: number of public and private hospitals in a city. Includes health centres. Crime rate: estimation of the general level of crime in a city. Slavery Index: the variable represents the national government's response to situations of slavery in the country. The countries that rank highest are the ones dealing with the problem most effectively. Happiness Index: countries with a higher value are those, where the level of overall happiness is higher. Gini Index: index values range from 0 to 100. A value of 0 expresses perfect equality of income distribution, and 100, maximal inequality. Global Peace Index: this index measures the level of peace/violence in a country or region. Countries with a high level of violence rank lowest. Health Care Index: estimation of the overall quality of the health care system, health care professionals, equipment, personnel, costs, etc. LGBT-friendly: this variable indicates whether a city provides a friendly environment for the LGBT community (on a scale of 1 to 5). Cities with a value of 1 have a more hostile environment for this community; those with a value of 5 are very LGBT-friendly. Price of property: property price as a proportion of income. Calculated as the ratio of the average price of a home to average annual disposable household income. Female employment rate: rate of female employment in the public sector. Value from 0 to 1. Death rate: death rate per 100,000 city inhabitants. Unemployment rate: unemployment rate (unemployed/labour force). Murder rate: murder rate per 100,000 city inhabitants. Suicide rate: suicide rate per 100,000 city inhabitants. Terrorism: number of terrorist incidents in a city in the last 3 years. Racial tolerance: index of racial tolerance in a city.

Source: www.iese.edu/media/research/pdfs/ST-0633-E.pdf.

Table 4.3 Dimensions and indicators included in the IESE Cities in Motion Index (Economy Indicators)

Dimension	*Indicators*
Economy indicators	Ease of starting a business: top positions in the ranking are held by cities that have a more favourable regulatory environment for setting up and operating a local business. Mortgage: mortgage as a percentage of income is the monthly mortgage cost as a proportion of household income (the lower the better). Motivation of individuals to undertake early-stage entrepreneurial activity: the percentage of opportunity-driven early-stage entrepreneurs divided by the percentage of necessity-driven early-stage entrepreneurs. Number of headquarters: number of headquarters of publicly traded companies. GDP: gross domestic product in millions of USD. Estimated GDP: projected GDP growth for the next year. GDP per capita: gross domestic product per capita. Purchasing power: purchasing power in buying goods and services in the city (based on the average salary), compared to that of New York City residents. If local purchasing power is 40, this means that inhabitants with an average salary can afford to buy 60% less goods and services than New York City residents with an average salary. Productivity Labour: productivity calculated as GDP/employed population (in thousands). Hourly wage in USD: hourly wage in the city (in USD). Time required to start a business: number of calendar days needed to complete the procedures to legally operate a business.

Source: www.iese.edu/media/research/pdfs/ST-0633-E.pdf.

(Wolniak and Jonek-Kowalska, 2022), which makes it possible to combine business and community interests.

The area related to urban governance includes many different indicators (Table 4.4). They illustrate the quality of services offered to residents by the city government. Thus, those related to urban democracy and the ability to use urban administration can be found among them. The ranking also assesses the level of corruption of authorities and officials. Another important aspect is the IT and telecommunications services offered to the urban community. A rather

Table 4.4 Dimensions and indicators included in the IESE Cities in Motion Index (Governance)

Dimension	*Indicators*
Governance	Bitcoin legal: whether or not Bitcoin is legal in the city.
	Whether or not the city has ISO 37120 certification. Certified cities are committed to improving urban services and quality of life. This variable is coded from 0 to 6.
	The highest value is assigned to the cities that have been certified for the longest time. A value of 0 is assigned to cities that are not certified.
	Government buildings: number of government buildings and premises in a city.
	Embassies: number of embassies in a city.
	Public sector employment: percentage of employed population working in public administration and defence; education; health; community, social and personal service activities; and other activities.
	E-Participation Index: this index supplements the E-Government Development Index (EGDI) and focuses on the use of online services to facilitate provision of information by governments to citizens ('e-information sharing'), interaction with stakeholders ('e-consultation'), and engagement in decision-making processes ('e-decision-making').
	Human Capital Index: the EGDI is a composite measure of 3 important dimensions of e-government: provision of online services, telecommunication connectivity and human capacity. This variable captures the human capacity component.
	Strength of Legal Rights Index: this index measures the degree, to which collateral and bankruptcy laws protect the rights of borrowers and lenders, and thus facilitate access to loans. The index ranges from 0 (low) to 12 (high), with higher scores indicating that these laws are better designed to expand access to credit.
	Telecommunication Infrastructure Index: the EGDI is a composite measure of 3 important dimensions of e-government: provision of online services, telecommunication connectivity and human capacity. This variable captures the development status of telecommunication infrastructure (by the government).
	Corruption Perceptions Index: countries with values close to 0 are perceived as very corrupt and those with values close to 100 are perceived as very transparent.

Table 4.4 (Continued)

Dimension	*Indicators*
	Online Service Index: the EGDI is a composite measure of 3 important dimensions of e-government: provision of online services, telecommunication connectivity and human capacity. This variable reflects the scope and quality of e-government services. Research offices: number of research and technology offices in a city. Open data platform: whether or not the city has an open data system. Democracy Index: the top-ranked countries are the ones considered most democratic. Economist Intelligence Unit. Reserves: total reserves in millions of current USD: city-level estimate according to population. Reserves per capita: reserves per capita in millions of current USD.

Source: www.iese.edu/media/research/pdfs/ST-0633-E.pdf.

unusual measure is the size of a city's financial reserves, which can illustrate city's resilience to economic crises. Interestingly, the Governance area also takes into account certification under the 37120 standard, which also performs a comprehensive verification of city structures. Thus, it can be concluded that the area of urban governance (Wielicka-Gańczarczyk and Jonek-Kowalska, 2023a) is very comprehensively analysed.

The ranking, in line with the areas of Smart City analysis described in the previous section, also includes the environmental aspect (Table 4.5). It mainly concerns air pollution. Also present are issues of access to water and the ability to manage municipal waste. An interesting measure is the city's exposure to climate risk, rarely included in Smart City analyses.

Another of the assessment areas concerns, readily described and exposed, urban mobility and transportation in the city (Table 4.6). This area assesses the availability of all possible forms of transportation, including green ones.

A far less classic area of Smart City assessment is urban planning, often included as part of urban management, and in the ranking analysed here it is a separate area of assessment (Table 4.7).

Table 4.5 Dimensions and indicators included in the IESE Cities in Motion Index (Environmental Indicators)

Dimension	*Indicators*
Environment indicators	CO_2 emissions: CO_2 emissions from the use of fossil fuels and the manufacture of cement. Measured in kilotonnes (kt). Methane emissions: methane emissions caused by human activities, such as agriculture and industrial methane production. Measured in kt of CO_2 equivalent. Environmental Performance Index: from 1 = poor to 100 = good. CO_2 Emission Index. Pollution Index. PM10: a measure of particles in the air with a diameter of less than 10 µm. Annual mean. PM2.5: a measure of particles in the air with a diameter of less than 2.5 µm. Annual mean. Percentage of population with access to water supply. Renewable water resource per capita. Solid waste; average amount of municipal solid waste generated annually per person (kg/year). Waste Management for Everyone Climate vulnerability: risk to the city due to climate change.

Source: www.iese.edu/media/research/pdfs/ST-0633-E.pdf.

According to the data in Table 4.7, urban planning considers issues related to housing, but also those related to the availability of electric car charging stations or bike-sharing. These are aspects less often included in a comprehensive analysis of smart urban infrastructure.

An equally infrequently assessed area is International Profile, which was also ranked in the IESE Cities in Motion Index (Table 4.8).

In this regard, we have a number of indicators related to transportation, hotel or food and beverage infrastructure, which can be indicators of a city's internationalization. Yet they should be considered rather reserved for large, globally recognized cities. For this reason, some might consider them exclusionary.

The last group of assessments relates to technology, which, for many Smart City researchers, is a key determinant of being Smart (Table 4.9).

According to the data in Table 4.9, the assessment of technology in the ranking analysed relates primarily to IT and ICT accessibility.

Table 4.6 Dimensions and indicators included in the IESE Cities in Motion Index (Mobility and Transportation)

Dimension	*Indicators*
Mobility and transportation	Bicycle rental: whether or not the city has a bicycle rental system. Moped rental: whether or not the city has a moped rental system. Scooter rental: whether or not the city has a scooter rental system. Bicycles per household: percentage of bicycles per household. Bike-sharing: shows automated services for public use of shared bicycles that provide transportation from place to place in a city. Indicator values range from 0 to 8 according to how developed the system is. Metro stations: number of metro stations in a city. Traffic Inefficiency Index: this index is an estimate of traffic inefficiencies. High values represent high driving inefficiencies, such as long travel times. Traffic Commute Time Index: an index based on the time it takes to commute to work (in minutes). Exponential Traffic Index: this index is estimated by considering time spent in traffic. It is assumed that travel time dissatisfaction increases exponentially beyond 25 minutes. Length of metro system: length of the metro system in a city. High-speed train: binary variable that shows whether the city has a high-speed train or not. Vehicles in the city: number of commercial vehicles in a city. Flights: number of inbound flights (air routes) in a city.

Source: www.iese.edu/media/research/pdfs/ST-0633-E.pdf.

The ranking also examines personal and professional interest in social media. Furthermore, it also assesses the innovativeness of cities. It should be noted that the technological aspect does not overwhelm other areas of evaluation and is certainly not dominant. The authors of the ranking expose 3 key aspects in the ranking, namely: human capital, social cohesion and urban governance. The assessment is, therefore, balanced and based on a humanistic rather than strictly technological approach, which undoubtedly deserves attention and appreciation.

Table 4.7 Dimensions and indicators included in the IESE Cities in Motion Index (Urban Planning)

Dimension	*Indicators*
Urban planning	Bike advance: whether or not a city has a bike-sharing system. Buildings: the number of completed buildings in a city. The count includes structures such as high-rises, towers and low-rise buildings, but excludes other miscellaneous structures and buildings of different statuses (under construction, proposed, etc.). Bicycle stations: bicycle station locations in a city. Electric charging stations: electric car charging points in a city. Number of people per household: average number of people per household. Percentage of the urban population with adequate sanitation services: percentage of the urban population that uses at least basic sanitation services – that is, improved sanitation facilities that are not shared with other households. Artificial intelligence (AI) projects: whether or not a city has AI projects. High-rises: percentage of buildings classified as high-rises – a high-rise is a multi-floored building of at least 12 stories or 35 m in height (115 feet).

Source: www.iese.edu/media/research/pdfs/ST-0633-E.pdf.

Table 4.8 Dimensions and indicators included in the IESE Cities in Motion Index (International Profile)

Dimension	*Indicators*
International profile	Number of passengers per airport: annual number of passengers per airport in thousands. Hotels: number of hotels per capita. Restaurant Price Index: the Restaurant Price Index compares the price of meals and drinks in restaurants and bars in a city to prices in New York City. McDonald's: number of McDonald's establishments in a city. Number of congresses and meetings: number of international congresses and meetings held in a city.

Source: www.iese.edu/media/research/pdfs/ST-0633-E.pdf.

Table 4.9 Dimensions and indicators included in the IESE Cities in Motion Index (Technology)

Dimension	*Indicators*
Technology	Mobile broadband: active mobile broadband subscriptions. Innovation Cities Index: the Innovation Cities Index (ICI) is a ranking of leading cities in innovation. Internet: percentage of households with internet access. LTE/WiMAX: percentage of the population covered by at least an LTE/WiMAX mobile network. Computers/PCs: percentage of households with a personal computer. Mobile phone penetration rate: number of mobile phones per 100 inhabitants. Social media: registered Twitter users in a city (in thousands of individuals) + number of registered LinkedIn members in the city. Broadband subscriptions: broadband subscriptions per 100 inhabitants. Telephony: percentage of households with some kind of telephone service. Internet speed: fixed-line internet speed in megabytes per second (country). Mobile speed: mobile speed in megabytes per second (country). WiFi hotspots: total number of WiFi hotspots.

Source: www.iese.edu/media/research/pdfs/ST-0633-E.pdf.

The second ranking selected for description is the Smart City Index, which was developed by the Institute for Management Development (IMD) and Singapore University for Technology and Design (SUTD) (Bris et al., 2019, 2020). Its design is based on the assessments of 120 residents of the surveyed cities, ensuring its bottom-up objectivity. The assessment takes into account the 2 pillars of infrastructure and technology. Also, the evaluated cities are divided into 4 groups due to the Human Development Index (HDI), which avoids the dominance of cities with a high degree of economic and civilization development. In the introduction to the characteristics of each city, a universal metric is also included, which, in addition to the HDI, includes:

- Number of residents;
- Life expectancy at birth;

- Expected years of schooling;
- Mean years of schooling;
- GNI per capita.

The indicators assessed by residents in the infrastructure and technology group are shown in Tables 4.10 and 4.11. These are simple formulations that residents will easily understand and quickly be able to assess. Nor is the assessment time-consuming, since there are fewer indicators than in the IESE Smart Cities in Motion described above.

Thus, the infrastructure area also assesses health (Wielicka-Gańczarczyk and Jonek-Kowalska, 2023b) and safety. Air pollution

Table 4.10 Dimensions and indicators included in the Smart City Index (Structures)

Dimension	*Indicators*
Health & safety	Basic sanitation meets the needs of the poorest areas Recycling services are satisfactory Public safety is not a problem Air pollution is not a problem Medical services provision is satisfactory Funding with the rent equal to 30% or less of a monthly salary is not a problem
Mobility	Traffic congestion is not a problem Public transport is satisfactory
Activities	Green spaces are satisfactory Cultural activities (shows, bars and museums) are satisfactory
Opportunities (work & school)	Employment finding services are readily available Most children have access to a good school Lifelong learning opportunities are provided by local institutions Businesses are creating new jobs Minorities are welcome
Governance	Information on local government decisions are easily accessible Corruption of the city officials is not an issue of concern Residents contribute to decision-making of local government Residents provide feedback on local government projects

Source: https://imd.cld.bz/IMD-Smart-City-Index-Report-20231/36/.

Table 4.11 Dimensions and indicators included in the Smart City Index (Technologies)

Dimension	*Indicators*
Health & safety	Online reporting on city maintenance problems provides a speedy solution A website or app allows residents to easily give away unwanted items Free public WiFi has improved access to city services CCTV cameras have made residents feel safer A website or App allows residents to effectively monitor air pollution Arranging medical appointments online has improved access to those services
Mobility	Car-sharing Apps have reduced congestion Apps that direct you to an available parking space have reduced journey time Bicycle rental has reduced congestion Online scheduling and ticket sales have made public transport easier to use The city provides information on traffic congestion through mobile phones
Activities	Online purchasing of tickets to shows and museums has made them easier to attend
Opportunities (work & school)	Online access to job listings has made it easier to find work IT skills are successfully taught in schools Online services provided by the city have made it easier to start a new business The current internet speed and reliability meet connectivity needs
Governance	Online public access to city finances reduces corruption Online voting has increased participation An online platform, where residents can propose ideas, has improved the quality of city life Processing identification documents online has reduced waiting time

Source: https://imd.cld.bz/IMD-Smart-City-Index-Report-20231/36/.

and municipal waste issues are also identified within this area. Traditionally, the area of mobility appears in the ranking, with an assessment of 2 key determinants of residents' quality of life, which are traffic (Othman et al., 2024) and the availability of public transportation. The ranking also assesses leisure activities in green areas and cultural institutions.

An interesting area of evaluation is the so-called Opportunities area, in which the city is analysed in terms of educational, employment and business opportunities, as well as acceptance of social diversity. The entire infrastructural assessment closes with an area related to urban order, which is evaluated primarily in terms of communication and information. In this context, there are indicators related to: availability of urban information, community participation and administrative corruption.

The same main areas appear in the technological evaluation group as in the infrastructural aspect (Table 4.11), but the specific indicators relate to the use and availability of urban technologies.

Thus, health and safety technologies are based on the use of: WiFi, websites, apps and cameras in the process of providing medical services, keeping residents safe and monitoring the environment.

Technological assessment of mobility, on the other hand, includes indicators for applications that identify urban traffic, enable sharing economy solutions (Jonek-Kowalska and Wolniak, 2022) and facilitate the use of urban transportation networks. Most of these are informational in nature.

In contrast, the solutions evaluated in the urban activities area refer primarily to the possibility of purchasing tickets to cultural institutions via the internet, which facilitates the use of public services by residents and allows the institutions themselves to plan the process of providing these services more effectively.

In the area concerning employment and education, IT and ICTs are evaluated in the context of online services that facilitate finding a job and starting one's own business. An important aspect is also the formation of IT skills and competencies in schools, which allows the assimilation of technological innovations and contributes to their regular dissemination.

In urban governance, the described ranking evaluates the use of innovative technologies for the provision of municipal public services, such as participation in elections, public participation, administrative services or information on the state of public finances.

It follows from the above description that the Smart City Index is based primarily on the technicized concept of the Smart City, in which the quality of life is improved mainly with the use of modern technology and the infrastructure for its use. Although the ranking also does not lack indicators relating to human capital, environmental protection or social cohesion. What is important and distinctive about this report is that it is based on a survey of public opinion, which allows the Smart City to be viewed from the perspective of a resident, and which balances the technicization of the presented assessment.

In addition to international rankings showing some arbitrariness in the selection of indicators for evaluating Smart Cities, ISO 37120 is also used in the evaluation of Smart City solutions. It was published on 15 May 2014 by the International Organization for Standardization. Its purpose was to standardize the assessment of Smart Cities, so that comparisons can be made over time and space. Such a solution is useful for both individual assessments and benchmarking between cities (Wolniak, 2019).

ISO 37120 evaluates 100 indicators, 46 of which are primary and 54 of which are auxiliary, additional indicators. These indicators are assigned to the following groups:

1. Economy;
2. Education;
3. Energy;
4. Environment;
5. Finance;
6. Managed organizational order;
7. Fire protection and combating the effects of natural disasters;
8. Health;
9. Recreation;
10. Security;
11. Social assistance;
12. Solid waste;
13. Telecommunication and innovations;
14. Transportation;
15. Urban planning;
16. Wastewater;
17. Water and sanitation services.

With their use, the level of provision of individual public services, and thus the quality of urban life, is assessed. The aforementioned groups of assessments are more detailed than the areas, in which Smart City development is traditionally described (presented in the previous chapters), but they, nevertheless, relate directly to such issues as

Smart Economy, Smart Mobility, Smart Environment, Smart People, Smart Living and Smart Governance (Bitkowska and Łabędzki, 2021; Augustyn, 2020; Sobol, 2017). Thus, they can be taken as a more insightful basis for evaluating efforts to create smart urban structures. A detailed list of the primary and auxiliary indicators demonstrated in the 37120 standard is presented in Tables 4.12–4.17 by the areas of Smart City creation described above.

According to the data in Table 4.12, the area related to the city's economy primarily emphasizes key macroeconomic indicators related to unemployment and entrepreneurship (Jonek-Kowalska, 2023). The city's financial condition, on the other hand, is assessed in terms of expenditures, revenues and budget balance. These are classic data, collected by municipal authorities, and therefore easy to calculate and present.

The standard, like most rankings, also addresses urban mobility issues (Alnsour et al., 2024; Table 4.13). This, in turn, is detailed in the standard to urban transportation and urban transit. Auxiliary indicators also analyse the availability of bicycle paths and air transportation. In the area related to telecommunications, the standardization of the urban assessment concerns access to the internet, mobile

Table 4.12 Areas and indicators included in the 37120 standard: Smart Economy

Area	*Primary indicators*	*Auxiliary indicators*
Economy	Urban unemployment rate Estimated value of commercial and industrial properties as a percentage of the total assessed value of all real properties Percentage of city residents living in poverty	Percentage of people having full-time employment Unemployment rate among young people Number of businesses per 100,000 residents Number of new partners per 100,000 residents per year
Finances	Debt service rate (debt service expenses as a percentage of the municipality's own income)	Financial expenditures as a percentage of all expenditures Own revenues as a percentage of all revenues Taxes collected as a percentage of taxes accrued

Source: 37120 standard.

Table 4.13 Areas and indicators included in the 37120 standard: Smart Mobility

Area	*Primary indicators*	*Auxiliary indicators*
Transport	Number of kilometres of high-capacity public transportation system per 100,000 residents Number of kilometres of public transportation system per 100,000 residents Annual number of public transportation trips per capita Number of passenger cars per capita	Percentage of commuters using a method of travel other than a passenger car Number of single-track motor vehicles per capita Kilometres of bicycle paths and lanes per 100,000 residents Fatalities due to traffic accidents per 100,000 residents Commercial air travel (number of commercial non-stop air destinations)
Telecommunication and innovations	Number of internet lines per thousand residents Number of cell phone lines per 100,000 residents	Number of fixed telephone lines per 100,000 residents

Source: (37120 standard).

and fixed-line telephony. This illustrates the city's communication and digital capabilities.

The environmental area typically includes pollution levels, waste management, as well as wastewater collection and treatment facilities (Table 4.14). Noise levels and biodiversity are assessed as well. This approach allows for a holistic view of air and water quality, as well as the extent, to which problems related to the growing amount of municipal waste are being addressed.

The Smart People area (Table 4.15) is directly attributable by standard to indicators related to primary, secondary and higher education, which illustrates the level of education of the urban community. Among the parameters assessed is also the availability of teachers in primary school. Unlike the rankings, however, the standard does not directly refer to issues related to the digitization of the urban

Table 4.14 Areas and indicators included in the 37120 standard: Smart Environment

Area	*Primary indicators*	*Auxiliary indicators*
Environment	Concentration of fine particulate matter (PM2.5) Concentration of particulate matter (PM10) Greenhouse gas emissions measured in tonnes per capita	Concentration of NO_2 (nitrogen dioxide) Concentration of SO_2 (sulphur dioxide) Concentration of O_3 (ozone) Noise Percentage changes in the number of native species
Solid waste	Percentage of city residents covered by regular solid waste collection service (households) Total amount of municipal solid waste collected per capita Percentage of the city's solid waste that is recycled	Percentage of the city's solid waste disposed of in a landfill that meets sanitary requirements Percentage of the city's solid waste disposed of in an incinerator Percentage of the city's solid waste that is incinerated outside Percentage of the city's solid waste disposed of in open landfills Percentage of the city's solid waste disposed of by other methods Amount of hazardous waste generated per capita (tons) Percentage of the city's hazardous waste that is recycled
Wastewater	Percentage of city residents with access to wastewater collection service Percentage of untreated city wastewater Percentage of the city's wastewater that has undergone first stage treatment Percentage of the city's wastewater that has undergone second stage treatment Percentage of the city's wastewater that has undergone third stage treatment	-

Source: 37120 standard.

Table 4.15 Areas and indicators included in the 37120 standard: Smart People

Area	*Primary indicators*	*Auxiliary indicators*
Education	Percentage of girls attending schools Percentage of students who completed primary school Percentage of students who graduated from secondary school Ratio of students to primary school teachers	Percentage of boys attending schools Percentage of adolescents attending schools Number of degrees obtained in higher education per 100,000 residents

Source: 37120 standard.

community (indicators from this area appear only within the framework of accessibility to telecommunications services).

Given that the standard directly addresses the quality of urban life, it devotes a lot of space to Smart Living (Table 4.16). This area consists of issues related to the supply and consumption of energy, including that from renewable sources. One can also find issues related to the health of residents and concerning the level and availability of health care (Jonek-Kowalska, 2022b). Overall quality of life is also identified in the context of life expectancy and mental health of residents. Finally, among additional quality parameters, one also finds those relating to recreational opportunities within and outside the city limits. Thus, the proposed approach takes into account both the level of satisfaction of essential existential needs, as well as additional options that significantly affect final satisfaction with quality of life.

According to the data in Table 4.17, the most areas included in the standard can be attributed to Smart Governance. These primarily concern the availability of public services and the satisfaction of public collective needs, as this is the area, in which the activities of the city government and their impact on the quality of life of the urban community are most often assessed. Interestingly, the area also includes indicators on urban democracy. They illustrate the level of interest of residents in the functioning of the city government, thus providing an assessment of the extent of citizen participation in the process of creating a Smart City.

Notably, in addition to the indicators described above, in assessing Smart Cities, the standard also assumes the development of an

Table 4.16 Areas and indicators included in the 37120 standard: Smart Living

Area	*Primary indicators*	*Auxiliary indicators*
Energy	Total household electricity consumption per capita (kWh/year) Percentage of city residents covered by legal electricity supply Energy consumption of public buildings per year (kWh/m^2) Percentage of total energy obtained from renewable sources, as part of total municipal energy consumption	Total energy consumption per capita (kWh/year) Average number of energy outages per customer per year Average length of energy outages (in hours)
Health	Average life expectancy Number of hospital beds per 100,000 residents Number of doctors per 100,000 residents Under-5 mortality rate per 1,000 live births	Number of nursing and midwifery personnel per 100,000 residents Number of mental health practitioners per 100,000 residents Percentage of suicides per 100,000 residents
Recreation	-	Number of square metres of public indoor recreational space per capita Number of square metres of public outdoor recreational space per capita

Source: 37120 standard.

introduction containing basic information about the city in question, as well as demographic and economic data. The latter are called profile indicators (Wróbel et al., 2023). The combination of general information and detailed indicators provides a holistic picture of a city in a uniform, standardized, quantified form.

The approach proposed in the standard allows for a transparent assessment of the quality of urban services, which ultimately translates into the main goal of Smart Cities, which is to improve the quality of life for current and future generations. Furthermore, the use of a number of mathematized indicators objectifies the assessment and

Table 4.17 Areas and indicators included in the 37120 standard: Smart Governance

Area	*Primary indicators*	*Auxiliary indicators*
Management	Voter turnout in recent local elections (as a percentage of of those eligible to vote) Percentage of women elected to office at the municipal public administration level, among all elected	Percentage of women employed in municipal administration Number of convictions for corruption and/or bribery of city representatives per 100,000 residents Representation of residents: number of local representatives elected to office per 100,000 residents Number of registered voters as a percentage of the number of residents of voting age
Spatial planning	Green areas (hectares) per 100,000 residents	Annual number of trees planted per 100,000 residents. Area of informal habitats as a percentage of the city's area Ratio of jobs to housing units
Social assistance	Percentage of city residents living in slums	Number of homeless per 100,000 residents Percentage of households without registered titles
Water and sanitation services	Percentage of city residents with access to potable water Percentage of city residents with access to water treated for consumption Percentage of residents with access to improved sanitation facilities Total water consumption per person in the household (litres/day)	Total water consumption per capita (litres/day) Average annual number of hours of water supply interruptions per household Percentage of water loss (missing water)
Safety	Number of police officers per 100,000 residents Number of crimes against life per 100,000 residents	Crimes against property per 100,000 residents Police response time from receipt of first call Percentage of violent crimes per 100,000 residents

(*Continued*)

Table 4.17 (Continued)

Area	*Primary indicators*	*Auxiliary indicators*
Fire protection and combating natural disaster effects	Number of fire-fighters per 100,000 residents Number of deaths due to fires per 100,000 residents Number of deaths due to natural disasters per 100,000 residents	Number of part-time and volunteer fire-fighters per 100,000 residents Response time of emergency services from receipt of first call Response time of fire department from receipt of first call

Source: 37120 standard.

enables precise analysis and accountability for the effects of decisions made by city authorities. And the international nature of ISO 37120 makes it possible for cities to systematically monitor their progress in innovating urban structures and striving to be smart (Wolniak, 2019).

In practice, the use of the described standard allows to gain a broader knowledge of the city's functioning locally and globally. This provides a basis for more effective city management and the provision of higher-quality public services. Quantitative analysis of city government performance can also provide a valuable source of knowledge for city stakeholders, including, in particular, the local community and investors. The use of this analysis promotes transparency and credibility that facilitates the process of making electoral or investment decisions. Interestingly, certification of a city's performance carried out based on the 37120 standard can be an additional asset in the process of applying for external funding in competitions for national and international projects (Fijałkowska and Aldea, 2017).

The aforementioned certification is based on the number of indicators monitored by the city and includes 5 levels of evaluation. The highest – Platinum – requires calculating and studying all 46 primary indicators and a minimum of 45 auxiliary indicators. The next 2, Gold and Silver, involve monitoring all primary indicators and a minimum of 30 and 14 auxiliary indicators, respectively. Obtaining the Bronze level requires calculating all primary indicators and any number of auxiliary indicators (as a last resort, one may not calculate any of them). For the last level, referred to as Aspirational, a minimum of 30 primary indicators must be monitored (Wróbel et al., 2023).

Certified cities are entered into the Global Cities Registry™. After a year, they go through the certification process again.

From the above description of the certification process, it is clear that the evaluation of Smart City solutions is based on the informational scope of the city data provided and monitored. The process does not value individual areas. Indeed, no boundaries or evaluation criteria are established. This approach makes it possible to create categorized rankings, which may be a drawback of the standard in question. However, the use of this tool in practice provides an opportunity to observe and benchmark Smart City development. Moreover, regardless of the results achieved, it gives all cities an equal chance to apply for the certificate, which is undoubtedly one of its advantages.

4.2 Assessment of areas and indicators of exclusion in selected Smart City rankings

Given the accusations against Smart Cities, it is worth analysing the previously described rankings and the 37120 standard in the context of capturing and evaluating possible pathologies related to urban exclusion. This analysis is carried out in this subsection divided into the 6 leading areas of Smart City formation and evaluation. In addition, the summary highlights other – non-area – possibilities for perceiving and studying the dimensions of exclusion in the Smart City.

The first ranking, described in the previous section, was the IESE Cities in Motion Index, and with it we begin to identify the measurement of exclusion. Table 4.18 shows the indicators that address the darker sides of the Smart City and the methods adopted to reveal them.

According to the data in Table 4.18, the IESE Smart City Index holistically addresses the risks of exclusion within social cohesion. There appear not only indicators related to economic exclusion, but especially parameters related to the risks of discrimination based on gender, age, sexual orientation and racial affiliation. This is undoubtedly a broad spectrum of analysis of possible exclusions. It is also worth adding that within the framework of economic exclusion, the Gini Index is taken into account, which not only relates to income issues, but reflects possible social inequalities.

In addition to the area relating directly to social cohesion, signalling of possible exclusions also appears in 5 other areas of evaluation. Thus, in the case of the Smart Economy, we deal with typical measures of income deprivation, such as GDP per capita and purchasing capacity.

Table 4.18 Exclusion dimensions identified in the IESE Smart City Index

Area	*Indicators referring to the risk of exclusion*	*Type of exclusion identified*
Social cohesion	Female-friendly Female employment rate Slavery Index Gini Index Unemployment rate LGBT-friendly Racial tolerance	Gender-based exclusion Economic exclusion Sexual orientation-based exclusion Race-based exclusion
Economy indicators	GDP per capita Purchasing power	Economic exclusion
Governance	Strength of Legal Rights Index Corruption Perceptions Index	Ethics-based exclusion
Environmental index	Percentage of population with access to water supply Waste Management for Everyone	Civilization- and environment-based exclusion
Urban planning	Number of people per household Percentage of the urban population with adequate sanitation services	Residential exclusion
Technology	Internet: percentage of households with internet access. Computers/PCs: percentage of households with a personal computer. Mobile phone penetration rate: number of mobile phones per 100 inhabitants	Digital exclusion

Source: www.iese.edu/media/research/pdfs/ST-0633-E.pdf.

In Governance, the ranking measures corruption in municipal government and administration. The environmental area, on the other hand, refers to the availability of drinking water and municipal waste disposal as civilization challenges in less developed areas. An interesting and far less common approach to measuring exclusion are the indicators of the number of people per household (Jonek-Kowalska, 2022c) and the availability of sanitation facilities present in the urban planning area. This approach allows for assessing the scale of potential housing shortages and identifying problems of overcrowding. Finally, in the technology area, there are indicators related to digital exclusion. They concern 3 dimensions: Internet, access to personal computers and cell phones.

Accordingly, the IESE Cities in Motion ranking makes it possible to identify the potential threat of exclusion in a very broad and multifaceted dimension. Its creators do not disregard, therefore, the problems and dark sides of the Smart City. With the help of the presented evaluation, it also becomes possible to identify the scale of possible threats and their benchmarking on a national and international scale.

The second ranking analysed is the Smart City Index based on the opinion of city residents. Table 4.19 lists indicators that identify different dimensions of exclusion for the 2 key areas considered in these rankings, namely Structures and Technology.

According to the data presented in Table 4.19, the ranking under review primarily considers digital exclusion. This is due to the focus of this assessment on the technical and infrastructural aspects of Smart City development, including internet accessibility, in particular.

Table 4.19 Exclusion dimensions identified in the Smart City Index

Area	*Indicators referring to the risk of exclusion*	*Type of exclusion identified*
Structures	Basic sanitation meets the needs of the poorest areas Funding with the rent equal to 30% or less of a monthly salary is not a problem Most children have access to a good school Minorities are welcome Corruption of the city officials is not an issue of concern	Economic exclusion Educational exclusion Exclusion based on nationality, race, sexual orientation, etc. Ethics-based exclusion
Technology	Free public WiFi has improved access to city services IT skills are successfully taught in schools The current internet speed and reliability meet connectivity needs Online public access to city finances reduces corruption	Digital exclusion Ethics-based exclusion

Source: https://imd.cld.bz/IMD-Smart-City-Index-Report-20231/36/.

Nevertheless, despite the relatively low number of indicators included in the overall assessment, the ranking also includes economic, educational and ethical exclusion. In addition, several types of exclusion can be included in the minority indicator, since this parameter does not specify the type of minorities tolerated. Thus, we can deal in this case with national minorities, as well as sexual or religious minorities. This approach is due to the synthetic nature of the assessments in the Smart City Index.

This monograph also examines the 37120 standard for assessing the quality of life and equivalence of cities. Thus, Table 4.20 shows the indicators referring to the problems of exclusion in this document.

Table 4.20 Exclusion dimensions identified in the 37120 standard

Area	*Indicators referring to the risk of exclusion*	*Type of exclusion identified*
Smart Economy	Percentage of city residents living in poverty Urban unemployment rate Unemployment rate among young people Debt service rate (debt service expenses as a percentage of the municipality's own income)	Economic exclusion at the level of the resident, the urban community and the city relative to other cities Possible age-based exclusion
Smart Mobility	Number of kilometres of public transportation system per 100,000 residents Fatalities due to traffic accidents per 100,000 residents Number of internet, mobile and landline connections per thousand residents	Communication exclusion Digital exclusion
Smart Environment	Percentage of city residents covered by regular solid waste collection service (households) Percentage of city residents with access to wastewater collection service Amount of hazardous waste generated per capita (tonnes) Percentage of the city's solid waste that is recycled	Environmental exclusion

Table 4.20 (Continued)

Area	*Indicators referring to the risk of exclusion*	*Type of exclusion identified*
Smart People	Percentage of girls attending schools Percentage of students who completed primary school Percentage of students who graduated from secondary school	Educational exclusion
Smart Living	Percentage of city residents covered by legal electricity supply Average number of energy outages per customer per year Average length of energy outages (in hours) Average life expectancy Number of hospital beds per 100,000 residents Number of doctors per 100,000 residents Under-5 mortality rate per 1,000 live births	Energy exclusion related to energy poverty and intermittent energy supply Exclusion from access to health care
Smart Governance	Percentage of women elected to office at the municipal public administration level, among all elected Percentage of women employed in municipal administration Number of convictions for corruption and/or bribery of city representatives per 100,000 residents Percentage of city residents living in slums Percentage of city residents with access to potable water Percentage of city residents with access to water treated for consumption Number of crimes against life per 100,000 residents Response time of emergency services from receipt of first call Response time of fire department from receipt of first call	Gender-based exclusion Ethics-based exclusion Economic exclusion Civilization exclusion Exclusion related to security risks

Source: Authors' own compilation based on the 37120 standard.

According to the data presented in Table 4.20, economic exclusion in the standard is identified through poverty and unemployment, which relates directly to the quality of life of the population, with a particular focus on youth. Therefore, it can be concluded that, according to the authors of the standard, this age group may be particularly exposed to the lack of opportunities for economic development, which is an important determinant of satisfaction with urban life.

In addition to exclusion at the individual level, the economic indicators included in the standard also refer to the financial situation of the city (debt service rate), which makes it possible to consider the level of the urban community and can provide a starting point for national and international analysis. In this context, the scale of exclusion can be assessed at the regional and national levels.

In the area of Smart Mobility, the identification of exclusion is done – as in the previously described rankings – with reference to transportation and digital exclusion. It is expressed, respectively, in terms of indicators of the availability and safety of public transportation, as well as the availability of the internet, as well as mobile and landline telephony. They are identified at the individual level and relate directly to the quality of life of residents and the community.

Quality of life also depends on environmental factors. Given the lack of basic facilities in this regard, one can speak of environmental exclusion referring, in the case of 37120 standard, to the provision of regular collection of municipal waste and sewage (Jonek-Kowalska, 2022a). An important aspect of such exclusion, projecting, in particular, the quality of life of future generations, is also the issue of the generation and handling (recycling) of waste, including hazardous waste, in particular.

The standard also identifies the dangers of educational exclusion. They can be analysed in the context of the level of schooling. The lower it is, the greater the danger associated with a lack of understanding or aversion to new technologies in the Smart City. Education also indirectly affects the creation of innovations, and thus the possibility of developing Smart City solutions.

In the area of Smart Living, the standard refers to several types of possible exclusions. Although Smart Cities are defined and perceived in the context of those units that offer an above-average standard of living, nevertheless, it is not always available to everyone. Moreover, not all cities in the world – due to different levels of civilizational and economic development – can offer such a standard even to selected residents. Thus, some residents and some cities may be affected by energy poverty, manifested by lack of access to energy or interruptions in its supply. Energy exclusion may also result from a lack of financial resources to cover the cost of electricity and heat.

The quality of urban life is also strongly dependent on access to and quality of medical services. These factors directly affect not only the well-being of residents, but also their life expectancy and mortality, especially among infants and children. Hence, the 37120 standard exposes indicators that assess the aforementioned aspects.

It is also worth mentioning that the level of medical care and its availability only partially depends on the city government, as it is influenced by political, macroeconomic and civilizational conditions specific to the economy. For these reasons, restrictions on access to medical services can generate exclusions both at the level of the Smart City community and at the level of Smart Cities themselves considered in national and international terms.

Within the framework of Smart Governance, a number of indicators relating to different types of exclusions can be found in the standard. Some of them can be treated as universal threats that apply to all cities, while others are of a nature directly related to the level of economic, social and civilizational development and apply, in particular, to those cities that operate in poorly performing economies. Universal exclusions include those relating to discrimination against women in urban government and administration or slums dwellers. In contrast, exclusions more broadly affecting countries with low levels of economic development include lack of continuous access to potable water, high crime rates or corruption. Although in cities of highly developed economies such pathologies that threaten the quality of urban life are also increasingly encountered.

The analysis shows that the 37120 standard on sustainable urban development treats the issue of exclusions somewhat more broadly than international rankings. The latter are more strongly oriented towards the assumption that a Smart City, or a city aspiring to the status of being smart, no longer has civilization problems and does not face pathologies. Nevertheless, it is worth noting that such an assumption may be unrealistic and assumes the exclusivity of Smart Cities in relation to other entities. Already such an assumption may be exclusionary. Moreover, the situation of cities around the world is constantly changing, and it cannot be assumed that some of the well-developed ones will not be affected by exclusionary problems in the future. Failure to monitor these risks means a lack of control over the risks, and this poses a serious threat to the quality of life in modern cities.

The analysis shows that all of the selected Smart City assessments address exclusion issues. Therefore, they cannot be accused of an idealistic approach to the concept of developing smart urban structures. The most extensive exclusions are identified in the IESE Cities in Motion Index and the 37120 standard. However, these are very extensive assessments based on more than 100 indicators. The Smart City Index is a less extensive ranking, and its potential shortcomings in assessing exclusions are offset by the evaluation's reliance on residents' opinions, which is quite rare.

The rankings described later in this monograph, namely IESE Cities in Motion Index and Smart City Index, will be used to present empirical results on the identification of risks associated with various forms of exclusion in the Smart City.

Bibliography

Alnsour, J., Arabeyyat, A. R., Al-Hyari, K., Al-Bazaiah, S. A. I., Aldweik, R. (2024). Enhancing city logistics for sustainable development in Jordan: A survey-based study. *Logistics*, 8, 1. https://doi.org/10.3390/logistics8010001.

Ang-Tan, R., Ang, S. (2022). Understanding the smart city race between Hong Kong and Singapore. *Public Money & Management*, 42(4), 231–240. https://doi.org/10.1080/09540962.2021.1903752.

Augustyn A. (2020). Zrównoważony rozwój miast w świetle idei Smart City. *Wydawnictwo Uniwersytetu w Białymstoku: Białystok*, 85, 84–85.

Banerjee, P. S., Mandal, S. N., De, D., Maiti, B. (2023). Quality of life modeling with invulnerable social-IoT using network structural entropy for sustainable smart city. *IETE Journal of Research*, 69(9), 5813–5821. https://doi.org/10.1080/03772063.2022.2160840.

Berrone, P., Ricart, J. E. (2022). IESE Cities in Motion Index. www.iese.edu/media/research/pdfs/ST-0633-E.pdf (accessed 20.05.2024).

Bitkowska, A., Łabędzki, K. (2021). Koncepcja inteligentnego miasta – definicje, założenia, obszary. *Polskie Wydawnictwo Ekonomiczne: Warszawa*, 2021, 5.

Bris, A., Cabolis, C., Lanvin, B., Caballero, J., Hediger, M., Milner, W., Jobin C., Pistis, M., Zargari, N. (2020). *IMD Smart City Index 2020*. www.imd.org/smart-city-observatory/home/(accessed 20.05.2024).

Bris, A., Cabolis, C., Lanvin, B., Caballero, J., Hediger, M., Milner, W., Jobin, C., Pistis, M., Zargari, N. (2019). *Smart City Index 2019*.

Caird, S. (2018). City approaches to smart city evaluation and reporting: case studies in the United Kingdom. *Urban Research & Practice*, 11(2), 159–179. https://doi.org/10.1080/17535069.2017.1317828.

Caird, S. P., Hallett, S. H. (2019). Towards evaluation design for smart city development. *Journal of Urban Design*, 24(2), 188–209. https://doi.org/10.1080/13574809.2018.1469402.

Caragliu, A., Del Bo, C. F. (2022). Smart cities and urban inequality. *Regional Studies*, 56(7), 1097–1112. https://doi.org/10.1080/00343404.2021.1984421.

Dashkevych, O., Portnov, B. A. (2023). Human-centric, sustainability-driven approach to ranking smart cities worldwide. *Technology in Society*, 74, 102296. https://doi.org/10.1016/j.techsoc.2023.102296.

Fijałkowska J., Aldea T. (2017). Raportowanie zrównoważonego rozwoju miast a norma ISO 37120. *Wydawnictwo Uniwersytetu Ekonomicznego we Wrocławiu: Wrocław*, 182.

He, L. (2023). Assessing the smart city: A review of metrics for performance assessment, risk assessment and construction ability assessment. *Cogent Economics & Finance*, 11(2). https://doi.org/10.1080/23322039.2023.2273651.

Jonek-Kowalska, I. (2023). The exclusiveness of smart cities: Myth or reality? Comparative analysis of selected economic and demographic conditions of Polish cities. *Smart Cities*, 6, 2722–2741. https://doi.org/10.3390/smartcities6050123.

Jonek-Kowalska, I. (2022a). Municipal waste management in Polish cities: Is it really smart? *Smart Cities*, 5, 1635–1654. https://doi.org/10.3390/smartcities5040083.

Jonek-Kowalska, I. (2022b). Health care in cities perceived as smart in the context of population aging: A record from Poland. *Smart Cities*, 5, 1267–1292. https://doi.org/10.3390/smartcities5040065.

Jonek-Kowalska, I. (2022c). Housing infrastructure as a determinant of quality of life in selected Polish smart cities. *Smart Cities*, 5, 924–946. https://doi.org/10.3390/smartcities5030046.

Jonek-Kowalska, I., Wolniak, R. (2022). Sharing economies' initiatives in municipal authorities' perspective: Research evidence from Poland in the context of smart cities' development. *Sustainability*, 14, 2064. https://doi.org/10.3390/su14042064.

Lee, J., Babcock, J., Pham, T. S., Bui, T. H., Kang, M. (2023). Smart city as a social transition towards inclusive development through technology: A tale of four smart cities. *International Journal of Urban Sciences*, 27(sup1), 75–100. https://doi.org/10.1080/12265934.2022.2074076.

Mokarrari, K. R., Torabi, S. A. (2022). Ranking cities based on their smartness level using MADM methods. *Sustainable Cities and Society*, 72, 103030. https://doi.org/10.1016/j.scs.2021.103030.

Mora, L., Deakin, M., Reid, A., Angelidou, M. (2019). How to overcome the dichotomous nature of smart city research: Proposed methodology and results of a pilot study. *Journal of Urban Technology*, 26(2), 89–128. https://doi.org/10.1080/10630732.2018.1525265.

Mora, L., Deakin, M., Zhang, X., Batty, M., de Jong, M., Santi, P., Appio, F. P. (2021). Assembling sustainable smart city transitions: An interdisciplinary theoretical perspective. *Journal of Urban Technology*, 28(1–2), 1–27. https://doi.org/10.1080/10630732.2020.1834831.

Othman, E., Cibilić, I., Poslončec-Petrić, V., Saadallah, D. (2024). Investigating noise mapping in cities to associate noise levels with sources of noise using crowdsourcing applications. *Urban Science*, 8, 13. https://doi.org/10.3390/urbansci8010013.

Parjanen, S., Rantala, T. (2021). Building an open innovation platform as a part of city renewal initiatives. *European Planning Studies*, 29(12), 2165–2183. https://doi.org/10.1080/09654313.2021.1903397.

Sobol A. (2017). Inteligentne miasta versus zrównoważone miasta. *Wydawnictwo Uniwersytetu Ekonomicznego w Katowicach: Katowice*, 78.

Wang, H. J. (2023). Smart city branding vision: multiple stakeholder perspectives. *Innovation: The European Journal of Social Science Research*, 1–25. https://doi.org/10.1080/13511610.2023.2296384.

Wielicka-Gańczarczyk, K., Jonek-Kowalska, I. (2023a). Perceptions and attitudes toward risks of city administration employees in the context of smart city management. *Smart Cities*, 6, 1325–1344. https://doi.org/10.3390/smartcities6030064.

Wielicka-Gańczarczyk, K., Jonek-Kowalska, I. (2023b). Involvement of local authorities in the protection of residents' health in the light of the smart city concept on the example of Polish cities. *Smart Cities*, 6, 744–763. https://doi.org/10.3390/smartcities6020036.

Wolniak R. (2019). Wykorzystanie normy ISO 37120 do zarządzania jakością życia w mieście. In *Wyzwania i uwarunkowania zarządzania inteligentnymi miastami*, Jonek-Kowalska I. (Ed.), Wydawnictwo Politechniki Śląskiej: Zabrze, 124.

Wolniak, R., Jonek-Kowalska, I. (2022). The creative services sector in Polish cities. *Journal of Open Innovation: Technology, Market, and Complexity*, 8, 17. https://doi.org/10.3390/joitmc8010017.

Wróbel, I., Bartosik, B., Gondek, P., Piwowar, B. (2023). Rozwiązania i wskaźniki transportowe w inteligentnych miastach – Część I. *Problemy Kolejnictwa*, 198, 71–74.

5 Exclusivity and inclusiveness of Smart City in practice

5.1 Leaders and losers of inclusiveness in Smart City

In this chapter, the considerations focus on the practical aspect of exclusion in Smart Cities. In the first part of the analysis, based on the IESE Cities in Motion Index, the 10 best and 10 worst cities were selected in terms of ratings included in the Social Cohesion area. This dimension of the assessment identified the most indicators related to the risk of exclusion. The results refer to 2023. Tables 5.1 and 5.2 and Figures 5.1 and 5.2 present lists of cities with their place in the ranking overall and in the social cohesion area.

The above material shows that the 10 best cities include units classified in the ranking up to the 70th place overall. This means that cities with a high level of Smart City maturity are characterized by a high level of inclusiveness. It can therefore be concluded that being smart also means being open to otherness and, thus, a low risk of exclusions of various types (Lee et al., 2022; Waghmare and Singhal, 2021; Laenens et al., 2019).

Nevertheless, it is interesting that cities with a high rating in terms of social cohesion are not at the top of the overall evaluation in the IESE Cities in Motion Index. One is from the top 10, 3 from the second 10, 2 from the third 10, 2 from the 10, 1 from the fifth 10 and 1 from the seventh 10. Given the above, there is a relationship between the places in the overall ranking, but it is not strong, which means that social goals can be pursued in a relatively independent manner from other values characteristic of a Smart City.

Analysing the geographical location of the 10 best cities, we can notice that as many as 6 of them are located in Europe, including 3 countries from the Scandinavian region. The next 4 are located in

DOI: 10.4324/9781003499992-6

Table 5.1 Top 10 cities for Social Cohesion in the IESE Cities in Motion Index

City – Country	*Social Cohesion position*	*Ranking position*
Taipei – Taiwan	1	34
Edinburgh – UK	2	35
Canberra – Australia	3	40
Copenhagen – Denmark	4	10
Wellington – New Zealand	5	70
Bern – Switzerland	6	28
Ottawa – Canada	7	50
Munich – Germany	8	11
Eindhoven – Netherlands	9	49
Helsinki – Finland	10	20

Source: www.iese.edu/media/research/pdfs/ST-0633-E.pdf

Table 5.2 Ten worst cities for Social Cohesion in the IESE Cities in Motion Index

City – Country	*Social Cohesion position*	*Ranking position*
Kyiv – Ukraine	173	117
Bogota – Colombia	174	132
Rio de Janeiro – Brazil	175	136
Cape Town – South Africa	176	141
San Salvador – El Salvador	177	157
Lagos – Nigeria	178	183
Athens – Greece	179	121
Johannesburg – South Africa	181	164
Karachi – Pakistan	182	182
Caracas – Venezuela	183	179

Source: www.iese.edu/media/research/pdfs/ST-0633-E.pdf

4 different areas of the world, namely, Asia, Australia, Oceania and North America. According to the assessment, Europe as a continent is therefore characterized by the lowest level of urban exclusion risk. Additionally, in terms of social cohesion, the Scandinavian countries seem to be the best place to live.

A distinguishing feature of social cohesion leaders is also a high level of civilization and economic development. The prepared list does not include cities with a low level of GDP per capita. There are

Figure 5.1 Top 10 cities for Social Cohesion and their place in the ranking of IESE Cities in Motion Index.

Source: Authors' own study.

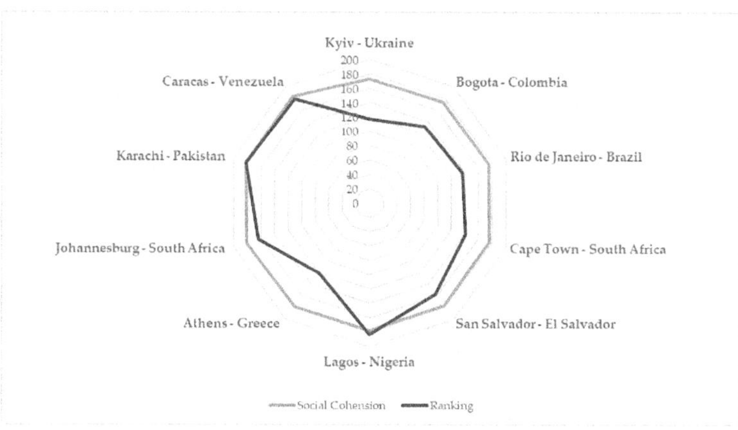

Figure 5.2 Ten worst cities for Social Cohesion and their place in the ranking of IESE Cities in Motion Index.

Source: Authors' own study.

no African cities in it either. Thus, it can be concluded that social well-being, centuries-old tradition, technological progress and a high level of education are factors that promote tolerance and social sensitivity to prevent exclusion. Given the nature of the indicators included in the area of social cohesion, this seems quite obvious. A high level of economic and civilizational development directly affects the availability of new technologies, education, health care or public safety (Mouton and Burns, 2021; Curtis, 2019).

Unfortunately, the above conclusions are not optimistic for many cities located in less developed regions (Khan, 2021; Evans, 2019). This is also reflected in the analysis of the 10 worst cities in terms of Social Cohesion in the IESE Cities in Motion Index. This is because these are the cities among the last 63 ranking units overall. Geographically, these are agglomerations located primarily in Africa, South America and Asia.

The countries where they are located are characterized by low levels of economic development, many economic and social difficulties and high levels of crime and corruption (Culver, 2021). Tensions along social and economic lines, as well as low levels of public security resulting from the above conditions, are a serious constraint on the social trust necessary for reducing various forms of exclusion (Plagerson, 2021). Notably, an analysis of the spread of the locations of the surveyed cities in the overall ranking versus in the area of social cohesion (Figures 5.1 and 5.2) shows that unfavourable economic-civilizational conditions are much more strongly correlated with the risk of exclusion than favourable conditions in this regard. In the case of the 10 worst cities, the described spread is much smaller than in the case of the 10 best cities.

Summarizing the above considerations, we can conclude that a high level of economic and civilizational development can be a stimulant of Smart City inclusiveness. On the other hand, a low level of the abovementioned factors is certainly a destimulant of inclusiveness and contributes to an increase in the threat of various forms of exclusion.

In the following section, the analysis is given to the second of the studied IMD Smart City Index rankings. For this purpose, the 5 best and worst ranked cities were selected and their ratings were examined in the categories assigned in Chapter 4 to identify urban exclusion. These included the following statements evaluated by residents:

- Basic sanitation meets the needs of the poorest areas;
- Funding with the rent equal to 30% or less of a monthly salary is not a problem;

- Most children have access to a good school;
- Minorities are welcome;
- Corruption of the city officials is not an issue of concern;
- Free public WiFi has improved access to city services;
- IT skills are taught well in schools;
- The current internet speed and reliability meet connectivity needs;
- Online public access to city finances reduces corruption (IMD Smart City Index 2023).

The scores for the aforementioned categories are shown in Figures 5.3–5.11, assuming the 0–100 scale adopted in the ranking.

Thus, the data presented in Figure 5.3 shows that people living in the poorest areas of a city are excluded from the ability to meet basic sanitation needs in both the highest-rated cities and those at the bottom of the ranking. However, the scale of this exclusion is more than double in cities located in less developed regions of the world. Indeed, in the case of this ranking, cities located in Europe, including Scandinavia and Switzerland, also lead the way.

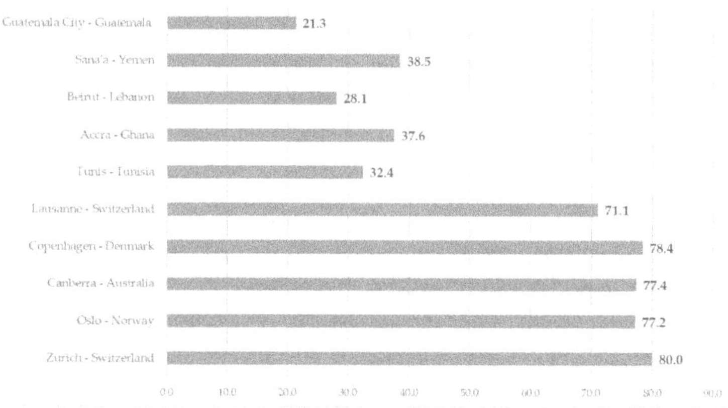

Figure 5.3 Basic sanitation meets the needs of the poorest areas: Residents' assessment of the five best and five worst cities in the IMD Smart City Index 2023.

Source: Authors' own work based on https://imd.cld.bz/IMD-Smart-City-Index-Report-20231.

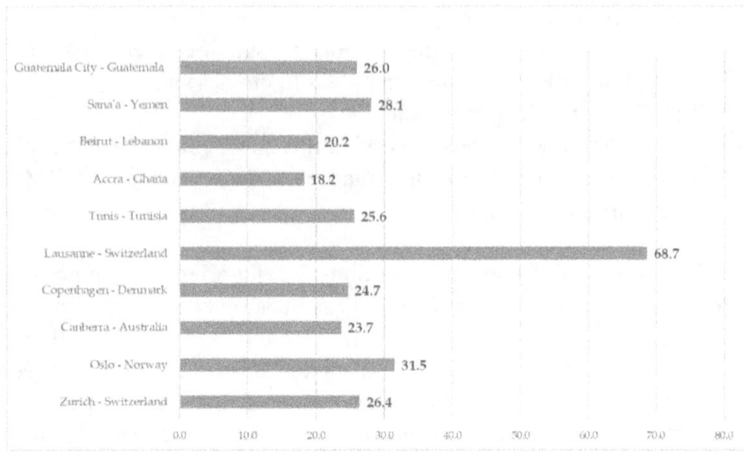

Figure 5.4 Funding with the rent equal to 30% or less of a monthly salary is not a problem: Residents' assessment of the five best and five worst cities in the IMD Smart City Index 2023.

Source: Authors' own work based on https://imd.cld.bz/IMD-Smart-City-Index-Report-20231.

Interestingly, in the second of the assessed categories related to economic and residential exclusion, almost all (with the exception of Switzerland's Lausanne) cities received very similar scores not exceeding 32% (Figure 5.4). This indicates a high level of housing exclusion risk due to the high value of the ratio of housing rents to the average monthly salary. In less economically developed cities this is due to low levels of wages, while in economically developed cities it is in turn due to high housing rents. Lausanne, on the other hand, which presents a deviation from the observed trends, is inhabited by very high-income residents, which has a favourable effect on the studied economic relationship, despite the high level of rents.

When it comes to assessing the accessibility of education, the cities at the top of the ranking again perform best (Figure 5.5), with little variation in their ratings, indicating that high Smart City status also means very good educational opportunities. For the cities with the worst positions in the IMD Smart City Index 2023, the ratings are significantly lower, indicating a high risk of educational exclusion. The ratings are also more diversified. It is worth mentioning, however,

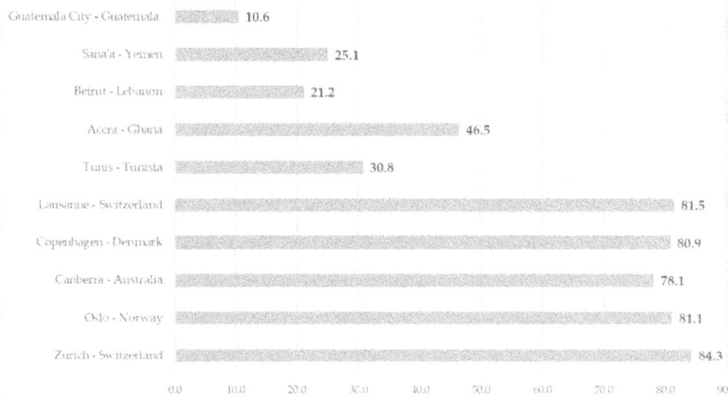

Figure 5.5 Most children have access to a good school: Residents' assessment of the five best and five worst cities in the IMD Smart City Index 2023.

Source: Authors' own work based on https://imd.cld.bz/IMD-Smart-City-Index-Report-20231.

that the ratings presented are subjective in nature, as they are based on the opinions of residents, who in less developed regions may have significantly lower expectations than residents of large European cities that are well developed in terms of civilization (Mimi et al., 2024; Abdelkarim, 2022). Consequently, the educational reality of the former may be worse than that ultimately shown by the ranking. The above observations confirm the strong dependence of the risks of exclusion on the economic situation of the region in which the cities aspiring to be smart are located.

Another area of evaluation refers to the acceptance of various types of minorities (Figure 5.6). What is characteristic of the described comparison of cities is the relatively small spread of ratings between the worst- and best-rated cities in the analysed ranking. It is the smallest among the surveyed responses, indicating that the threat of exclusion of minorities (sexual, religious, ethnic, etc.) is an issue for both leaders and outsiders of the Smart City rankings (Zhou et al., 2023; van Gils and Bailey, 2021). The reported data also shows that minorities are best accepted in Australia's Canberra. In European cities,

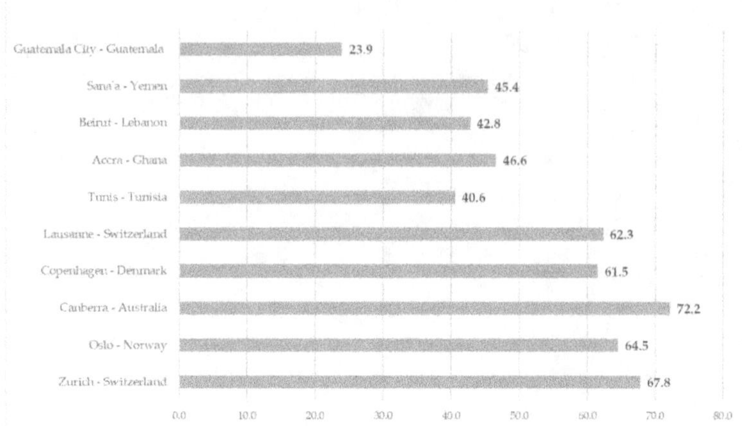

Figure 5.6 Minorities are welcome: Residents' assessment of the five best and five worst cities in the IMD Smart City Index 2023.

Source: Authors' own work based on https://imd.cld.bz/IMD-Smart-City-Index-Report-20231.

minorities are less readily accepted. This may be due to the increased migration of people from Africa to Europe, particularly intense in the last decade (Andersson Nystedt et al., 2024; Dumont, 2015). The problems associated with this phenomenon and the accompanying clashes between supporters and opponents of migration raise public concerns and arouse resentment against national and ethnic minorities, which certainly has an impact on the ultimate willingness of European city dwellers to accept 'otherness' (Giglioli, 2019; Wyss, 2019). Thus, it can be concluded that Smart Cities are not as open as optimistically assumed in the social aspect of the Smart City concept (Marchesani et al., 2024; Shin et al., 2024).

Another exclusionary threat analysed relates to corruption at the level of the city government and the level of residents' concern about the existence of this phenomenon (Figure 5.7). Thus, the ratings of residents presented below allow 2 general conclusions to be drawn. In the worst ranked cities, the threat is very high, as the rating of this aspect is dramatically low. In practice, therefore, the activities of the city government are of concern to residents, and the process of providing public services is non-transparent, and therefore limited in

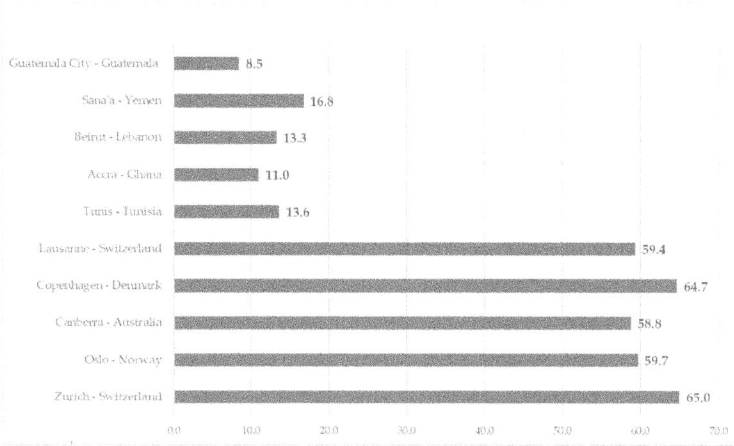

Figure 5.7 Corruption of the city officials is not an issue of concern: Residents' assessment of the five best and five worst cities in the IMD Smart City Index 2023.

Source: Authors' own work based on https://imd.cld.bz/IMD-Smart-City-Index-Report-20231.

quantity and quality. The urban community is therefore unable to take full advantage of a key resource for a Smart City, which is public administration.

Interestingly, corruption risk ratings are significantly higher in the cities at the top of the ranking, however, only in Copenhagen and Zurich do they exceed 60%. This means that even the smartest urban units are not free from ethical risks and exclusion of residents due to the lack of means, methods or personal contacts used to influence municipal decisions. This is a disturbing phenomenon that certainly requires deeper investigation and answers to the question of how much of the subjective opinions of residents are confirmed by actual incidents, and how much are the result of circulating opinions that 'authorities steal'.

The next 4 aspects of exclusion selected from the IMD Smart City Index relate to technology, including primarily those online. Figure 5.8 shows an assessment of internet accessibility in the provision of public services. According to the data presented, in both the best and worst cities, access to public services via WiFi is not

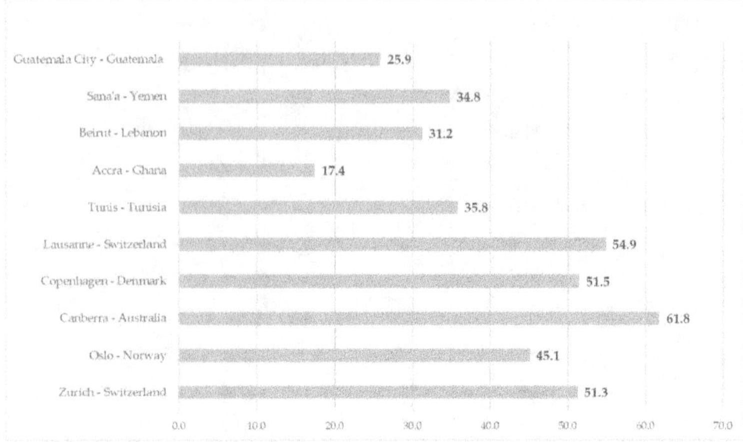

Figure 5.8 Free public WiFi has improved access to city services: Residents' assessment of the five best and five worst cities in the IMD Smart City Index 2023.

Source: Authors' own work based on https://imd.cld.bz/IMD-Smart-City-Index-Report-20231.

widespread and rated highest. Therefore, city authorities are not able to guarantee this desirable contemporary technology in a complete and effective way, despite the fact that technological development is one of the key attributes of a Smart City (Park and Yoo, 2022; Voordijk and Dorrestijn, 2019). Meanwhile, for the majority of residents – both in the more and less economically developed regions – improving the quality of urban governance is mainly related to the digitization of administrative processes. The need to strengthen the digitization of public services is therefore one of the important challenges facing Smart City authorities in both less and more developed regions of the world.

The lack of very clear differentiation of ratings in the top and bottom cities of the IMD Smart City Index is also evident in the area concerning the teaching of IT skills. In 8 out of 10 units analysed, this aspect's rating exceeds 40% (Figure 5.9). Only in Sana'a and Guatemala City is the residents' rating significantly lower. Of course, this is a subjective satisfaction, since the educational process is evaluated from the residents' perspective. Nonetheless, it is worth

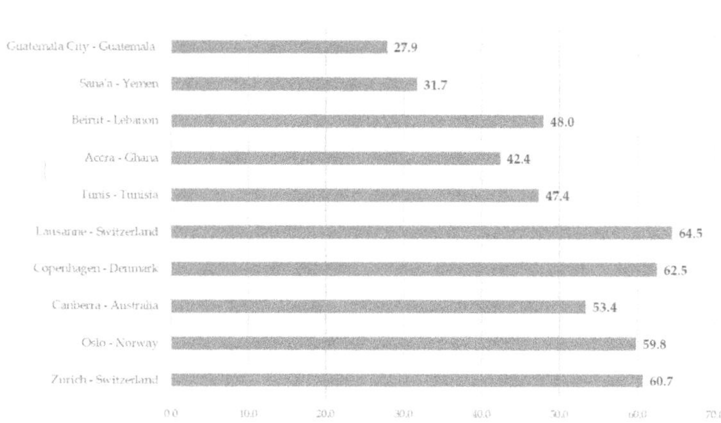

Figure 5.9 IT skills are taught well in schools: Residents' assessment of the five best and five worst cities in the IMD Smart City Index 2023.

Source: Authors' own work based on https://imd.cld.bz/IMD-Smart-City-Index-Report-20231.

noting that the modern world is digitally globalized, so verification of IT skills can be done easily and on the same basis regardless of geographic location. Still, however, access to digital technologies itself remains differentiated.

It's also worth noting that residents of highly developed regions with top ranking cities can certainly have very high expectations in terms of IT education. This, in turn, may result in a wider gap between expectations and reality and affect the low rating of this area (from 59% to 65%) by the local community in Lausanne, Copenhagen, Canberra, Oslo or Zurich.

An even lower spread between the best and worst cities in the ranking characterizes the fulfilment of expectations in terms of internet speed and reliability (Figure 5.10). In this case, some cities from less developed regions come close to the leaders of the described list. Such an assessment clearly indicates the dominance of the importance of internet technologies in modern cities regardless of their geographical location.

Therefore, are the surveyed cities threatened by digital exclusion understood as the lack of satisfactory access to the internet? The

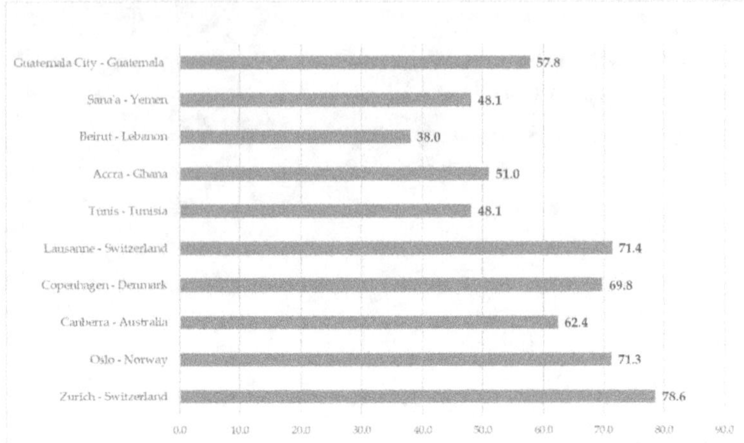

Figure 5.10 Current internet speed and reliability meet connectivity needs: Residents' assessment of the five best and five worst cities in the IMD Smart City Index 2023.

Source: Authors' own work based on https://imd.cld.bz/IMD-Smart-City-Index-Report-20231.

scores obtained, in all cases higher than 38%, indicate that it is not an aspect as badly assessed as the threat of corruption in city government. Thus, it can be concluded that technology providers perform far better than municipal decision-makers, although certainly the evaluation of internet speed and reliability is less complex than the evaluation of many activities and decisions made at the level of municipal administration.

It is worth noting, however, that even the leaders of the surveyed ranking do not guarantee 100% internet availability, making the threat of digital exclusion an estimated missing 20% In less developed cities, the gap ranges from 23% to 72%, so the possibility of being digitally excluded in these regions is significantly higher.

The last aspect of exclusion considered links corruption issues with online access to public finances (Figure 5.11). The function of this connection – as intended – is to increase the transparency of financial decisions of the city government contributing to the reduction of corrupt behaviour (Korosteleva et al., 2020; Cheng et al.,

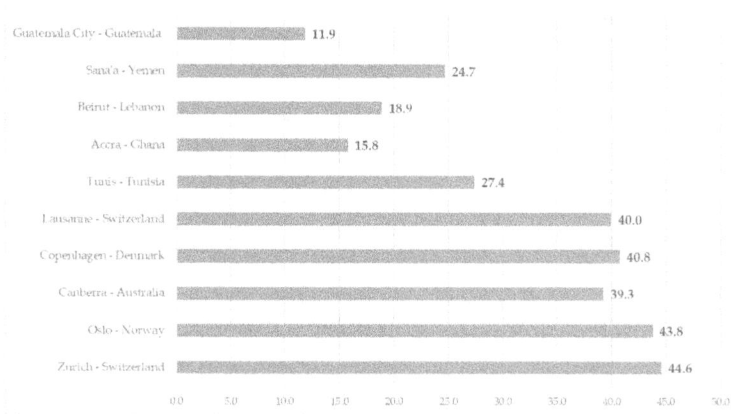

Figure 5.11 Online public access to city finances reduces corruption: Residents' assessment of the five best and five worst cities in the IMD Smart City Index 2023.

Source: Authors' own work based on https://imd.cld.bz/IMD-Smart-City-Index-Report-20231.

2023; Jones, 2019). It also performs a classic informational function, allowing the community to learn about the directions of investment in the city.

Thus, from the evaluations received by the authors of the ranking, it appears that the effects on the effects of the link identified above are less than satisfactory in both groups of compared cities (a rating of less than 50%). At the same time, the cities at the end of the ranking assess the extent of the reduction in corruption due to the provision of online financial data far worse than the leaders of the described list. This difference is the result of 2 circumstances. The first relates to the very high extent of corruption in less developed cities. The second relates to the speed and reliability of the internet, which was rated worse in the cities in the last places of the ranking than in the leading units. Regardless of the circumstances, however, it is worth noting that the authorities of the surveyed cities have a lot of work to do both in the transparency of financial and investment decisions and in the area of corruption reduction.

Taking everything into account, from the comparative analysis of the scale of exclusion in the top 5 and 5 worst cities of the IMD Smart City Index, the following conclusions can be drawn:

- The differences in the risk of exclusion in the technology area between the cities at the top and the bottom of the ranking are significantly smaller than those observed in the structures area;
- The largest differences between the analysed groups are observed for: (1) economic exclusion relating to access to sanitation for the poorest, (2) educational exclusion relating to the availability of good schools, and ethical exclusion identified by the level of corruption in city government;
- Interestingly, the differences regarding digitization are not as significant as those mentioned above, moreover, in the opinion of residents, digital exclusion does not seem to be as significant a threat as the economic or ethical one;
- A major challenge for both groups surveyed is the issue of minority acceptance and corruption in city government.

5.2 Smart City inclusiveness best practices: Case studies

This section focuses on case studies of cities described in the literature in the context of inclusion and, therefore, exemplifying best practices to prevent various exclusions. The case studies described were developed based on a review of the literature on the subject, and the results of this review are included in Table 5.3.

The literature on urban exclusion is dominated by cases describing the image and policy of inclusiveness. Cities thus promote their openness and friendliness. This approach has the character of a top-down presentation of inclusiveness and minimization of the risks associated with urban exclusion.

Relatively few studies focus on a bottom-up assessment of inclusiveness conducted among residents or those at risk of exclusion, which can distort a proper assessment of actual social cohesion, which, as the previously cited ranking assessments show, even in the highest-rated Smart Cities is not satisfactory. Bottom-up surveys are more difficult to carry out and are fraught with subjectivity, but nevertheless allow the opinion of the local community, which plays a key role in the Smart City concept, to be known.

Table 5.3 Case studies on best practices for Smart City inclusivity

City	*Dimensions of inclusiveness*
Rotterdam (Netherlands)	High and accepted cultural and ethnic diversity – more than half of residents have an immigrant background (Belabas, 2023; Dukes and Musterd, 2012). High tolerance for diverse minorities that is part of the city's image and as a value also shared by residents (Belabas, 2023). Plenty of green space conducive to communal recreation regardless of weather conditions – initiating and strengthening social ties (Belabas, 2023; Casais and Monteiro, 2019; Stedman et al., 2004).
Dubai (United Arab Emirates)	Emphasizing inclusivity in the city's image concerning age and disability. Combining traditional and modern (Alsayel et al., 2022).
Amsterdam (Netherlands)	Emphasizing in the city's image inclusivity in terms of: religion and ideology, race and ethnicity, gender and sexual orientation. Offering residents plenty of green recreational areas (Alsayel et al., 2022). Bottom-up and top-down urban solidarity with migrants. Having an office for undocumented migrants where they can obtain housing for 18 months (Özdemir, 2022).
Toronto (Canada)	Emphasizing in the city's image inclusivity in terms of: religion and ideology, race and ethnicity, gender and sexual orientation. Offering residents a range of international cultural events (Alsayel et al., 2022).
Berlin (Germany)	Bottom-up and top-down urban solidarity with migrants. Having an official city policy on city services for migrants. Offering migrants an anonymous and free health card to guarantee access to medical care (Özdemir, 2022).
Barcelona (Spain)	Bottom-up and top-down urban solidarity with migrants. Having an official municipal policy: housing, health and regarding other municipal services for migrants. Organizing assistance for migrants in the form of jobs, subsidies and moral support at the community level, as well as demonstrations in support of accepting refugees (Özdemir, 2022).
Melbourne (Australia)	Conducting workshops for children and youth with disabilities to identify mobility issues and improve the functioning of the city's disability-friendly infrastructure (Rachele et al., 2024).

(*Continued*)

Table 5.3 (Continued)

City	*Dimensions of inclusiveness*
Beijing (China)	Combining urban planning with the formation of social cohesion by taking into account not only urban infrastructure, but also its connection to socio-cultural events. Such an approach can foster the inclusion of minorities, especially migrants, in shared urban life (Wang and Liu, 2022).
Bristol, Cardiff, Glasgow, Liverpool, Peterborough (UK)	Developing and implementing The Inclusive Cities programme, which aims to collaborate between universities and city governments to increase the integration of migrants into the local community. The authors expose the following measures for increasing social cohesion: increasing the role of local leadership, carving out spaces for integration and developing horizontal networking (Broadhead, 2020).
Vienna (Austria)	Implementing the Smart City Wien 2050 strategy, which emphasizes the role of social innovation in enhancing social cohesion. Urban decision-makers also declare their willingness to identify diversity in its various forms and protect it by offering a high level of public safety. They also guarantee access to local services of the highest standard for all residents (Romanelli, 2022).
Florence (Italy)	The city has developed the Firenze Smart City Plan (2015), which includes putting residents first and seeking shared spaces that foster collaboration and inclusivity. The city's activities aim to promote: integration, innovation, involvement and information (Romanelli, 2022).
Surakarta (Indonesia)	The city became a rehabilitation centre for people after the Civil War (1946–1950), which increased the visibility of disabilities and the integration of people with disabilities into the community. The city is also developing highly accessible infrastructure for people with disabilities. At the national level, a number of pieces of legislation aimed at improving the quality of life and integration of people with disabilities have also been developed (Patrick and McKinnon, 2022).
San Pietro Mussolino (Italy)	This small city is an example of the successful integration of 20% migrants from Mexico, Ghana and India. The migrants assimilated into the community mainly by securing jobs in the local industry of marble mining and running tanneries (De Vita and Oppido, 2016).

Table 5.3 (Continued)

City	*Dimensions of inclusiveness*
Xiong'an	The city was treated as a laboratory to test ways to increase social cohesion in various areas. To this end, a state-of-the-art medical and educational centre was built there. Mobility was also strengthened with high-speed rail. Farmers were offered an investment programme. Furthermore, the area of forested land was increased from 11% to 40% (Liu, 2020).

Source: Authors' own work based on publications cited in the table.

Also, they make it possible to confront the assumptions of urban inclusivity policies with the effects of their application. For example, Belabas and George (2023), describing Rotterdam's inclusive policy, emphasize that it is not effective in practice. Indeed, their bottom-up research on a group of young migrants shows that the image promoted by the city does not encourage migrants to identify with the urban community. Nor does it reduce their sense of otherness or alienation.

A certain seemliness of Rotterdam's efforts at inclusivity is also pointed out by Schiller et al. (2023). They examine the impact of partnerships between different organizations on real social cohesion and the inclusion of migrants and others at risk of exclusion in urban life. Their findings show that Rotterdam has a layered network of many organizations working for inclusiveness, but only a few of them play an active role in reducing the threat of exclusion.

Very often, the city's overall very good evaluation of inclusivity is not confirmed in all areas and aspects of its functioning. Such a situation is described by some researchers, including Kristensen et al. (2023) who use the example of mobility in Copenhagen (Denmark). The authors note that in the mobility situation in different neighbourhoods of the city differs from each other and significantly so. This causes residents of less well connected areas of the city to feel excluded in terms of mobility. Summary or resultant opinions on Smart Cities are therefore not always fully reflected in the opinions of all city stakeholders.

The discussion so far also shows that the topic of exclusion appears in the literature in fragmented terms. This is because it deals with certain types of exclusion that are attractive and topical at a given time. Thus, there are many publications right now on the exclusion of migrants, while far less attention is paid to exclusion due to age, gender, disability or sexual orientation. There is also a lack of studies that comprehensively consider issues of urban exclusion and ways to counteract these phenomena. For these reasons, further considerations in this monograph will be a certain proposal to systematize the issues of urban exclusion and an attempt to create a framework for assessing the scale of this phenomenon.

Bibliography

Abdelkarim, S. (2022). The role of economic conditions in shaping citizens' satisfaction with government in Tunisia's fragile democracy: An expectancy-disconfirmation analysis. *International Journal of Public Administration*, 46(15), 1047–1060. https://doi.org/10.1080/01900692.2022.2069117.

Alsayel, A., de Jong, M., Fransen, J. (2022). Can creative cities be inclusive too? How do Dubai, Amsterdam and Toronto navigate the tensions between creativity and inclusiveness in their adoption of city brands and policy initiatives? *Cities*, 128, 103786. https://doi.org/10.1016/j.cities.2022.103786.

Andersson Nystedt, T., Herder, T., Agardh, A., Asamoah, B. O. (2024). No evidence, no problem? A critical interpretive synthesis of the vulnerabilities to and experiences of sexual violence among young migrants in Europe. *Global Health Action*, 17(1). https://doi.org/10.1080/16549716.2024.2340114.

Belabas, W. (2023). Glamour or sham? Residents' perceptions of city branding in a superdiverse city: The case of Rotterdam. *Cities*, 137, 104322. https://doi.org/10.1016/j.cities.2023.104323.

Belabas, W., George, B. (2023). Do inclusive city branding and political othering affect migrants' identification? Experimental evidence. *Cities*, 133, 104119. https://doi.org/10.1016/j.cities.2022.104119.

Broadhead, J. (2020). Building inclusive cities: Reflections from a knowledge exchange on the inclusion of newcomers by UK local authorities. *CMS*, 8, 14. https://doi.org/10.1186/s40878-020-0172-0.

Casais, B., Monteiro, P. (2019). Residents' involvement in city brand co-creation and their perceptions of city brand identity: A case study in Porto. *Place Branding and Public Diplomacy*, 15, 229–237. https://doi.org/10.1057/s41254-019-00132-8.

Cheng, M., Meng, Y., Jin, J., Nainar, K. (2023). Poverty mitigation and anti-corruption campaigns: Evidence from Chinese cities. *Journal of the Asia Pacific Economy*, 1–31. https://doi.org/10.1080/13547860.2023.2178160.

Culver, C. (2021). Chinese investment and corruption in Africa. *Journal of Chinese Economic and Business Studies*, 19(2), 119–145. https://doi.org/10.1080/14765284.2021.1925822.

Curtis, S. (2019). Global cities as market civilisation. *Global Society*, 33(4), 437–461. https://doi.org/10.1080/13600826.2019.1577805.

De Vita, G. E., Oppido, S. (2016). Inclusive cities for intercultural communities. *European Experiences. Procedia – Social and Behavioral Sciences*, 223, 134–140. https://doi.org/10.1016/j.sbspro.2016.05.333.

Dukes, T., Musterd, S. (2012). Towards social cohesion: Bridging national integration rhetoric and local practice: The case of the Netherlands. *Urban Studies*, 49(9), 1981–1997.

Dumont, A. (2015). Muslim Moroccan migrants in Europe: Transnational migration in its multiplicity. *Ethnic and Racial Studies*, 39(3), 505–507. https://doi.org/10.1080/01419870.2015.1095314.

Evans, J., Karvonen, A., Luque-Ayala, A., Martin, C., McCormick, K., Raven, R., Palgan, Y. V. (2019). Smart and sustainable cities? Pipedreams, practicalities and possibilities. *Local Environment*, 24(7), 557–564. https://doi.org/10.1080/13549839.2019.1624701.

Giglioli, I. (2019). On not being European enough: Migration, crisis and precarious livelihoods on the periphery of Europe. *Social & Cultural Geography*, 22(5), 725–744. https://doi.org/10.1080/14649365.2019.1601248.

Jones, P. (2019). Urban governance and its disorders: Corruption in the cities. *International Journal of Regional and Local History*, 14(2), 55–61. https://doi.org/10.1080/20514530.2019.1673540.

Khan, S. (2021). Barriers of big data analytics for smart cities development: A context of emerging economies. *International Journal of Management Science and Engineering Management*, 17(2), 123–131. https://doi.org/10.1080/17509653.2021.1997662.

Korosteleva, J., Mickiewicz, T., Stępień-Baig, P. (2020). It takes two to tango: Complementarity of bonding and bridging trust in alleviating corruption in cities. *Regional Studies*, 54(6), 851–862. https://doi.org/10.1080/00343404.2019.1652894.

Kristensen, N. G., Lindberg, M. R., Freudendal-Pedersen, M. (2023). Urban mobility injustice and imagined sociospatial differences in cities. *Cities*, 137, 104320. https://doi.org/10.1016/j.cities.2023.104320.

Laenens, W., Mariën, I., Walravens, N. (2019). Participatory action research for the development of e-inclusive smart cities. *Architecture and Culture*, 7(3), 457–471. https://doi.org/10.1080/20507828.2019.1679447.

Lee, J., Babcock, J., Pham, T. S., Bui, T. H., Kang, M. (2022). Smart city as a social transition towards inclusive development through technology: A tale of four smart cities. *International Journal of Urban Sciences*, 27(sup1), 75–100. https://doi.org/10.1080/12265934.2022.2074076.

Liu, Z., de Jong, M., Li, F., Brand, N., Hertogh, M., Dong, L. (2020). Towards developing a new model for inclusive cities in China: The case of Xiong'an new area. *Sustainability*, 12, 6195. https://doi.org/10.3390/su12156195.

Marchesani, F., Masciarelli, F., Bikfalvi, A. (2024). Exploring the (dis) advantages of smart cities' inclusive, integrative and social practices in new business creation: The effect of human capital inflow. *Entrepreneurship & Regional Development*, 1–31. https://doi.org/10.1080/08985626.2024.2358969.

Mimi, M. B., Kibria, Md. G., Selim, Md. M. I. (2024). How do economic, health, environmental and demographic factors affect life expectancy? A novel attempt for developed and developing economies. *International Journal of Sustainable Development & World Ecology*, 1–16. https://doi.org/10.1080/13504509.2024.2335269.

Mouton, M., Burns, R. (2021). (Digital) neo-colonialism in the smart city. *Regional Studies*, 55(12), 1890–1901. https://doi.org/10.1080/00343404.2021.1915974.

Özdemir, G. S. (2022). Urban solidarity typology: A comparison of European cities since the crisis of refuge in 2015. *Cities*, 130, 103976. https://doi.org/10.1016/j.cities.2022.103976.

Patrick, M., McKinnon, I. (2022). Co-creating inclusive public spaces: Learnings from four global case studies on inclusive cities. *Journal of Public Space*, 7(2), 93–116. https://doi.org/10.32891/jps.v7i2.1500.

Park, J., Yoo, S. (2022). Evolution of the smart city: Three extensions to governance, sustainability, and decent urbanisation from an ICT-based urban solution. *International Journal of Urban Sciences*, 27(sup1), 10–28. https://doi.org/10.1080/12265934.2022.2110143.

Plagerson, S. (2021). Mainstreaming poverty, inequality and social exclusion: A systematic assessment of public policy in South Africa. *Development Southern Africa*, 40(1), 191–207. https://doi.org/10.1080/0376835X.2021.1993793.

Rachele, J., Burn, G., Burke, K., Alisic, E. (2024). Improving inclusion for children and young people with a disability in inner-city Melbourne, Australia. *Cities*, 150, 105103. https://doi.org/10.1016/j.cities.2024.105103.

Romanelli, M. (2022). Towards smart inclusive cities. *PuntOorg International Journal*, 7(2), 216–234. https://doi.org/10.19245/25.05.pij.7.2.6.

Schiller, M, Awad, I., Buijse, N., Chantre, M., Huang, Y.-C., Jonitz, E., van den Brink, L., van Dordrecht, L. (2023). Brokerage in urban networks on diversity and inclusion: The case of Rotterdam. *Cities*, 135, 104219. https://doi.org/10.1016/j.cities.2022.104119.

Shin, J. E., Joe, M., Kim, E., Kang, N., Lee, J., Lee, S. W. (2024). Unboxing the blackbox of human-centric and inclusive smart city initiatives: A theory based evaluation framework. *Asian Journal of Technology Innovation*, 1–24. https://doi.org/10.1080/19761597.2024.2367762.

Stedman, R., Beckley, T., Wallace, S., Ambard, M. (2004). A picture and 1000 words: Using resident-employed photography to understand attachment to high amenity places. *Journal of Leisure Research*, 36(4), 580–606. https://doi.org/10.1080/00222216.2004.11950037.

van Gils, B. A. M., Bailey, A. (2021). Revisiting inclusion in smart cities: Infrastructural hybridization and the institutionalization of citizen participation in Bengaluru's peripheries. *International Journal of Urban Sciences*, 27(sup1), 29–49. https://doi.org/10.1080/12265934.2021.1938640.

Voordijk, H., Dorrestijn, S. (2019). Smart city technologies and figures of technical mediation. *Urban Research & Practice*, 14(1), 1–26. https://doi.org/10.1080/17535069.2019.1634141.

Waghmare, M., Singhal, S. (2021). Monitoring and evaluation framework for inclusive smart cities in India. *Development in Practice*, *32*(2), 144–162. https://doi.org/10.1080/09614524.2021.1907535.

Wang, X., Liu, Z. (2022). Neighborhood environments and inclusive cities: An empirical study of local residents' attitudes toward migrant social integration in Beijing, China. *Landscape and Urban Planning*, 226, 104495. https://doi.org/10.1016/j.landurbplan.2022.104495.

Wyss, A. (2019). Stuck in mobility? Interrupted journeys of migrants with precarious legal status in Europe. *Journal of Immigrant & Refugee Studies*, 17(1), 77–93. https://doi.org/10.1080/15562948.2018.1514091.

Zhou, S., Loiacono, E. T., Kordzadeh, N. (2023). Smart cities for people with disabilities: A systematic literature review and future research directions. *European Journal of Information Systems*, 1–18. https://doi.org/10.1080/0960085X.2023.2297974.

6 Proposal for assessing the inclusiveness of Smart City

6.1 Areas and principles of measuring inclusiveness

The considerations presented so far show that the subject of exclusions generated by Smart Cities appears both in the literature on the subject and in rankings and standards related to the assessment of Smart City infrastructure. However, it is not coherent, and exclusions are written and analysed in a fragmentary manner in various places. With the above in mind, in this chapter the authors have attempted to comprehensively arrange the knowledge on exclusion and inclusiveness in Smart Cities, combined with the presentation of a holistic concept of assessing the risk associated with urban exclusion.

It is worth starting to organize the issues related to urban exclusion and inclusiveness by defining these concepts and identifying their key determinants. Thus, **exclusion** in general means the inability of a resident to participate in activities undertaken and available to other representatives of the urban community (Richardson and Le Grand, 2002). At the same time, this inability is not due to the will of the individual, but to external circumstances, that constitute a barrier limiting accessibility, over which the resident in question has no control, and which they cannot influence.

In the above context, the source of exclusion is the urban system (Smart City system), which determines the position occupied by the resident and causes some to benefit from the many products, services and activities provided or made available by the city, while others systemically have such access restricted or prevented (Reimer, 2004; Szopa and Szopa, 2011). Thus, the result of a specified vision of the organization of the urban system is a kind of "handicap", resulting

DOI: 10.4324/9781003499992-7

in the deprivation of some residents of full participation in the urban community (paraphrasing the considerations of Giddens, 2004; Silver, 1994 on the general context of social exclusion). Given the multiplicity of products and services offered by urban systems, exclusion can, therefore, have very different causes and a multidimensional character.

It is also worth mentioning, that a Smart City, which is characterized by above-average technological, social and economic innovation, offers residents many new products, services or activities. Thus, the number of potential sources of risk of exclusion in various forms is increasing, and special attention should be paid to this. Accordingly, the Smart City system can also become a source of above-average **exclusion risk**. This risk, however, does not always have to materialize, since, by definition, it is only the possibility of a given risk and its associated loss (Knight, 1964; Haimes, 2015). Indeed, risks can be managed, and thus any risks associated with their occurrence can be minimized or offset (Pritchard, 2014; Terje, 2015). City authorities and stakeholders involved in the creation of urban systems and their infrastructure should play an active role in this process. With effective measures to manage the risk of exclusion, it will be possible to **strengthen the city's resilience to the risks** of various forms of exclusion, thereby increasing its inclusiveness.

Inclusiveness is, therefore, a marker of such an urban environment, in which all residents have the same access to products and services offered by the city. Then, they are not in any way threatened by the risk of exclusion and the consequences of realizing this risk. Issues of inclusiveness are currently being paid attention to in many dimensions, including primarily economic and social (Aliyev, 2021; Nizam et al., 2020), thus indigenously touching also on the issues described in this monograph from an urban perspective (Pięta-Kanurska, 2019).

In light of the above, inclusiveness is the opposite of exclusion. However, while exclusion – in very different forms and scope – can be encountered in practice, the full form of inclusiveness is idealistic in nature, and can only be a point to aim for in order to develop Smart Cities in a sustainable manner. One can, of course, assume that the vision of inclusiveness of Smart Cities is Utopian in nature (de Castro Neto and de Melo Cartaxo, 2021; Dalprà, 2020), but such an approach often has negative overtones and suggests the discontinuation of Smart City structures, which is no longer possible at their current stage of development. Efforts can only be made to improve them and make them more inclusive, and thus also sustainable in a multidimensional way.

The interrelationships between the concepts described above are shown in Figure 6.1. In Chapter 7, they will be used as a starting point for developing guidelines for strategies to strengthen the inclusiveness of Smart Cities.

In addition to defining the interrelationship of the concepts described in this monograph, it is worth taking a closer look at the types of exclusion that can be generated by the urban system. They are arranged in Table 6.1.

The developed classification of urban exclusion was based on the literature review and analysed ranking, as well as the 37120 urban standard. It was also supplemented with those exclusions that the authors believe may occur in Smart Cities and have not been reflected in previous studies and other sources.

The types of exclusions presented in Table 6.1 are arranged according to the hierarchy of urban needs, starting with the basic ones and ending with those, that can be treated as higher-order needs. This is a reference to the well-known theory of public and social goods in public finance (Holcombe, 1997; Samuelson and Marks, 2009) and Maslow's psychological theory of needs (Bridgman et al., 2018; Maslow, 1943).

Thus, the classification begins with civilizational and economic exclusions, which refer to meeting the basic needs of the urban community. Of course, they have long been known and analysed, not only at the urban level, but also, or perhaps primarily, at the level of individual economies. Thus, in the urban environment, they are most often derived from the civilizational and economic development of a country. Nevertheless, as the analyses presented in Chapter 5 show, the level of this development affects the scale of the exclusions mentioned and the final position of a given city in the Smart City rankings. Therefore, we can conclude that the creation of Smart City structures without taking into account civilizational or economic conditions can promote the deepening of differences between Smart Cities and the communities that inhabit them.

According to the authors, in the case of a high level of civilizational or economic exclusion, a city cannot be considered smart, because it does not meet the basic conditions for development and does not guarantee a minimum quality of life for a significant part of the local community. Indeed, this quality and its improvement is the foundation of

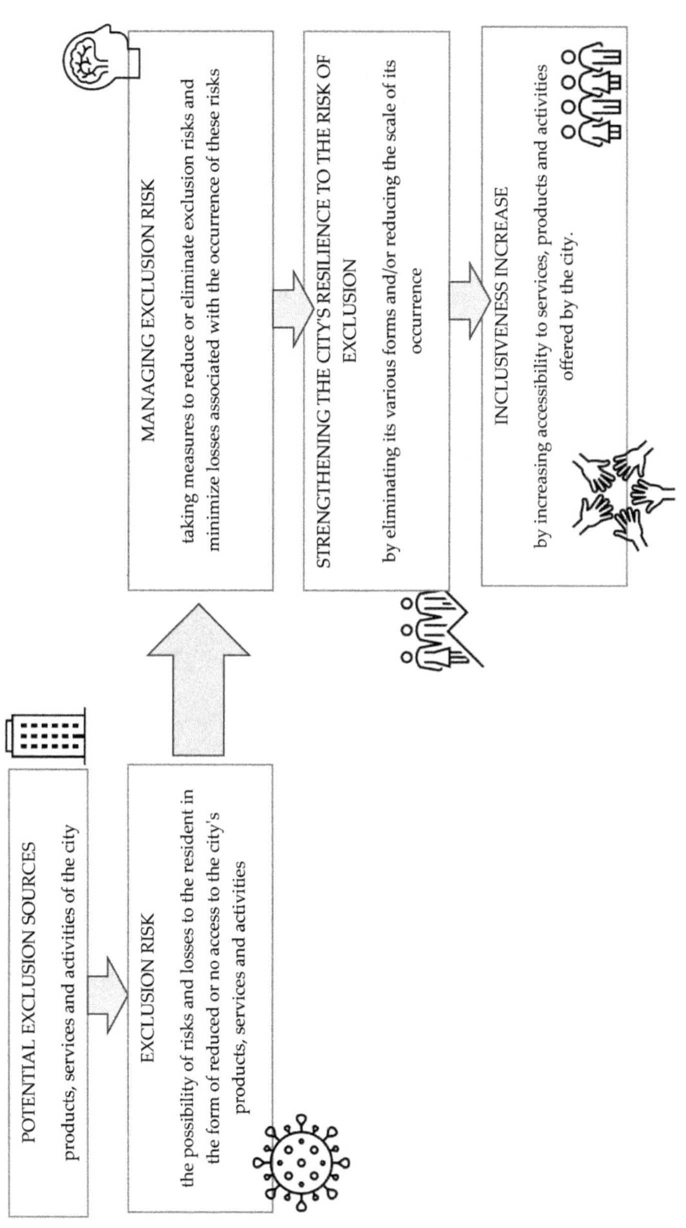

Figure 6.1 Relationship between exclusion and inclusiveness in Smart Cities.

Source: Authors' own study.

Table 6.1 Types of exclusion in Smart Cities

Type of exclusion	*Characteristics*
Civilizational	Relating to the availability of basic infrastructure, including primarily running water, electricity, sanitation, and reflecting the level of community poverty, including energy poverty.
Economic	Directly applicable to the property and income situation of the resident, including, in particular, aspects related to: • Employment (labour market situation, unemployment rate); • Housing (availability of housing, price of housing, level of rents, equipment of housing); • Disposable income (wage level, social security system). Indirectly applicable to the city's property and income situation, including, in particular: • The city's revenue capacity and financial self-reliance; • The level of the budget deficit; • The city's public debt and methods of financing it.
Public safety	Applicable to the threat of crime and the availability of preventive services (police, municipal police, sanitation services, etc.) and the extent of corruption in city government.
Educational	Applicable to the availability and quality of primary, secondary and higher education schools, and the level of community schooling.
Health	Applicable to the availability and quality of health care (primary care, hospitals, doctors, nurses, vaccinations, medicines) and the use of modern technology in health care.
Transportation	Applicable to accessibility and quality of roads, local public transport, rail and air transport. Also taking into account transportation sustainability, including the use of low-carbon modes of transportation or bicycle transportation.
Social	Directly applicable to the human rights to life, liberty and safety of a person, freedom of conscience and religion, freedom of expression, trial, citizenship, privacy, free elections, freedom of assembly and association, and the media. In particular, relating to non-discrimination on the basis of gender, age, sexual orientation, religion, race or disability.

Table 6.1 (Continued)

Type of exclusion	*Characteristics*
Environmental	Applicable to accessibility to waste management, wastewater treatment and recycling, as well as green spaces. Also taking into account the quality of the urban ecosystem, including levels of pollution of water, air and soil.
Cultural and sports	Applicable to the availability of cultural institutions and sports centres.
Digital	Applicable to the availability of computers, landline and cell phones, as well as the Internet. Also taking into account education for acquiring IT competencies and the availability of e-services at the local government level.

Source: Authors' own study.

the Smart City concept. Ensuring civilizational and economic needs in the basic scope can, therefore, be treated as a *sine qua non* condition for being smart in any other scope.

Notably, the level of income of a city and its residents – whether we like it or not – determines the availability of many other attributes of Smart Cities (modern technologies, the Internet, low-carbon transportation, etc.). It, therefore, has a direct impact on the availability and quality of urban infrastructure, and thus the quality of life of the local community. Confirmation of these observations is undoubtedly provided by cities at the top of the Smart City rankings, which operate in highly developed economies.

After exclusion of a civilizational and economic nature, the rankings show exclusion related to public safety. Indeed, in the rankings, standard and publications analysed, a lot of attention is paid to issues related to protecting citizens from mass threats and corruption as a common social phenomenon. Public safety is also one of the key public needs. The issue of safety is also included in Maslow's pyramid of needs.

Accordingly, the availability of services, that guarantee public safety, is an important determinant of the quality of urban life. The need for such safety is particularly acute during times of crises and threats, as could be seen in the case of the Covid-19 pandemic (Hassankhani et al., 2021; Yang and Chong, 2021; Sharifi et al., 2021).

City residents would also like to feel protected from the effects of corruption, which results in the exclusion or reduction of the availability of urban services on an equal basis for all (Doshi and Ranganathan, 2016). Meanwhile, as the analysed rankings show, discrimination against residents as a result of corruption occurs, and to a fairly large extent, not only in less developed economies, but also in well and perfectly functioning ones.

Educational and health issues were placed next in the ranking of possible exclusions. These can affect differences in quality of life both at the level of specific local communities living in a given city, as well as disparities that exist between cities located in different regions and countries. Education and health care are social goods that, depending on political doctrine, can be publicly and/or privately funded. This can cause limitations in their availability and equality of usage rules. Nevertheless, it is worth noting that their availability and quality are essential to the quality of urban life. It should also be added that urban health issues receive relatively little attention in analyses and studies (Wielicka-Gańczarczyk and Jonek-Kowalska, 2023). Meanwhile, the availability of health care and the quality of its services can be a source of significant exclusions in a Smart City, by definition promoting young and healthy people as representative residents (Wawer et al., 2022).

Among the exclusions of an infrastructural nature, those relating to mobility, which receives a great deal of attention in the context of a Smart City, have also been singled out. These exclusions can be analysed at the individual level, seeking answers to the question of whether all residents can use the available road and transportation infrastructure to the same extent and on the same terms? They can also be considered from a comparative perspective across regions and countries. In the context of higher-order mobility needs in a Smart City, it is certainly also worth looking at differences in the availability of sustainable transportation (Zawieska and Pieriegud, 2018).

Social exclusions, like those of mobility nature, often appear in the Smart City literature, because they directly affect individuals and communities. On the one hand, they are related to basic human rights (Kempin Reuter, 2019), and on the other, unfortunately, they are strongly characterized by individual views and beliefs, which are difficult to influence and affect. Therefore, they cause numerous controversies. The exclusion of refugees has received particular attention in recent years, due to the increasing processes of migration to large, well-developed cities. Exclusion on the basis of age (Bleja et al., 2020) or disability (Kolotouchkina et al., 2022) remains on the sidelines of consideration, the former of which, due to aging populations, is already likely to become a challenge of similar magnitude and intensity to population migration in the near future. The group of social exclusions also includes aspects related to the satisfaction of higher-order needs, such as the need to belong, relating to participation and urban democracy (van Gils and Bailey, 2023; Pereira et al., 2017).

The last 3 types of exclusions listed in the table can be treated in an elitist manner. This is because, on the one hand, they relate to the typical hallmarks of today's well-developed Smart Cities (modern technology, high living standards, work–life balance, clean environment), and on the other hand, they belong to higher-order needs, that ultimately determine the high quality of urban life. Due to economic, political or social problems (the exclusions described and mentioned earlier), many cities and the individuals or communities living in them cannot afford the luxury of being environmentally, culturally or digitally smart. That is because it is a higher stage in the development of both infrastructure and urban communities. This is especially true in terms of access to IT or ICT, as well as opportunities for cultural and sports entertainment.

In the case of environmental exclusion, one can speak of 2 layers of it. The first relates to accessibility to: waste management, wastewater treatment and recycling, as well as green spaces. The second also takes into account the quality of the urban ecosystem, including the level of pollution of water, air and soil. Again, from the perspective of highly developed cities and their well-to-do residents, the availability of the aforementioned elements seems obvious. Unfortunately, for many individuals and communities, it is a luxury good, from the use of which they are excluded (Scott, 2016).

The conceptual and classification apparatus described in this chapter will hereafter be used in the process of developing concepts for assessing the scale of exclusion and strategies for enhancing the invisibility of Smart Cities.

6.2 Concept of the Smart City inclusiveness assessment

Considering the presented way of capturing inclusiveness in the literature, rankings and studies, the proposal for assessing this phenomenon used a comprehensive approach trying to take into account all identified and classified forms of exclusion from the urban community. In this way, it sought to avoid the piecemeal nature of the assessment and the design approach to creating a Smart City, so often criticized by researchers and practitioners (Rosati and Conti, 2016; Caird and Hallett, 2018).

The analysis shows that exclusionary threats can refer to classic, already recognized forms of exclusion related to meeting basic individual and community needs (Willis, 2019). They can also relate to higher-order needs and refer to threats directly related to the creation of smart urban structures. For this reason, the assessment includes necessary and sufficient conditions for being inclusive (Kim, 2023; Gangadharan, 2021).

In addition, the proposed concept of assessing the inclusiveness of modern cities also involves a combination of top-down and bottom-up approaches. Most rankings, surveys, and analyses conducted by cities mainly take into account the measurement of inclusiveness made from the perspective of the city government, the creators of the ranking or researchers of a particular form of exclusion. However, the results of such assessments – which are admittedly based on statistical data – can be characterized by subjectivity in the selection and interpretation of indicators. They also do not take into account the opinions of the main urban stakeholders, for whom Smart Cities are being created.

Given the above, the authors of the study propose combining a top-down indicator assessment with a bottom-up assessment conducted from the perspective of a resident. This combination will allow for a more comprehensive view of urban exclusion and allow for confronting the results of city government and community assessment. The essence of this concept is shown in Figure 6.2.

Measuring inclusiveness begins by defining the dimensions taking into account the forms of exclusion described in this chapter. In this context, it is important to take into account all possible manifestations of exclusion, so that the initial analysis will provide data on their scale and scope of occurrence. In the subsequent stages related to the creation of a strategy for strengthening inclusiveness (described in Chapter 7), this will enable the separation of threats that are significant from those that do not occur in a given city or their intensity is so low, as to not be a problem.

Thus, bottom-up and top-down assessments are assumed to be carried out in parallel. The first requires the selection and assignment of appropriate measurement indicators, and at 2 levels: (1) basic

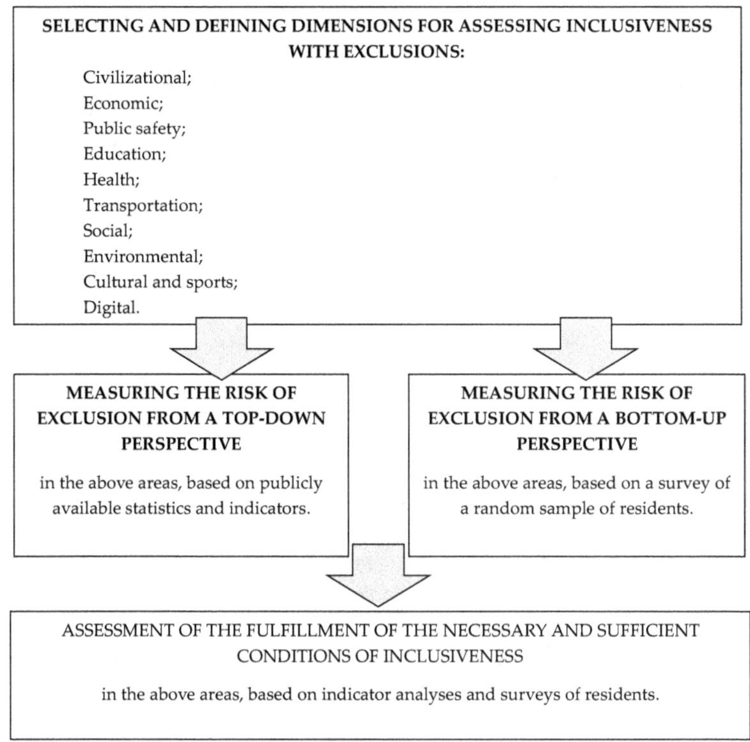

Figure 6.2 The concept of assessing inclusiveness of Smart Cities.

Source: Authors' own study.

(necessary conditions of inclusiveness), referring to the key determinants of satisfactory quality of life in the city; and (2) advanced (sufficient conditions of inclusiveness), referring to the city's satisfaction of higher-order needs related to the creation of smart urban structures. In selecting and assigning indicators to each assessment area, the following principles should be followed:

- Consideration of all types of exclusions;
- Assignment of several indicators to each area and level of exclusion;
- Universality, accessibility and comparability of the data used during the assessment (this will enable monitoring of indicators over time and space).

In the process of selecting measurement indicators, proposals contained in the Smart City rankings and the 37120 standard can be used. At the same time, those in the standard most often refer to the basic level of inclusiveness assessment, while those in the rankings refer to the advanced level of assessment directly related to the idea of Smart City. A proposal for selecting indicators at the first and second level is also presented later in this subsection.

The proposal to assess inclusiveness – in addition to a top-down indicator price – also assumes a bottom-up measurement conducted on the basis of residents' opinions. Such an approach may seem cumbersome, but it is nevertheless necessary, because without knowing the community's assessment, reliable identification of the risks of exclusion is not possible. After all, it should be remembered that Smart Cities are designed to serve residents and are created to improve their quality of life.

It is worth referring, at this point, to the principles of selecting a research sample. Naturally, its size should guarantee statistical representativeness. For cities with a population of 50,000 to 1 million, this means that about 380–400 residents need to fill out surveys (with a significance level of 5%, a fraction of 50% and an acceptable error of 5%). This is, therefore, not an impossible task or very time-consuming. It would be a good idea to stratify the structure of such a calculated sample according to the structure of the population of a given city, which means maintaining proportions in terms of such socio-demographic characteristics as gender, age, education, income level, etc. Maintaining the above principles ensures that the results obtained are representative, which is important in the process of generalizing them to the entire local community.

Survey questions, on the other hand, should relate directly to the particular types of risk of a given exclusion and correspond to the indicator assessment. For readability and transparency of the assessment, it is worth using a system of statements rated on a uniform scale, such as a 5-point Likert scale. This will guarantee ease, transparency and readability of the results, and respondents will facilitate the process of answering.

After collecting data and respondents' answers, the results of the bottom-up and top-down evaluation should be confronted with each other according to the procedure algorithm given in Figure 6.3. If the results of the top-down and bottom-up evaluation do not match, the reason for the differences should be identified. Once it is identified, a

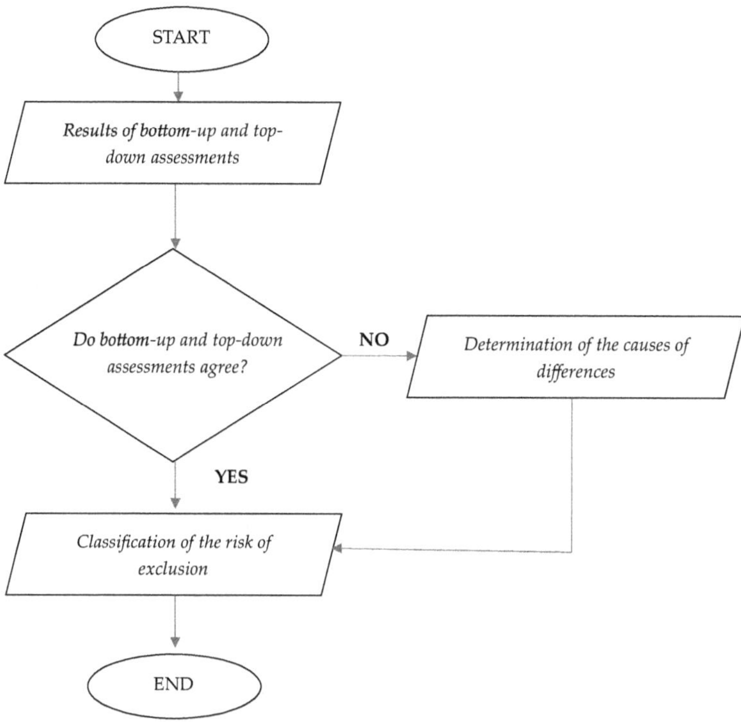

Figure 6.3 Algorithm for confronting bottom-up and top-down assessments of inclusiveness.

Source: Authors' own study.

decision can be made to categorize the exclusion threat as significant or insignificant (included in the risk mitigation process or not included).

If assessments are equal, the decision-maker only has to decide whether the threat of a given exclusion is significant or not, and thus, whether or not to plan mitigation measures against it. The process of assessing the scale of the exclusion risk threat itself is presented in Chapter 7.

An important element of the 2 inclusiveness assessments presented above is undoubtedly the selection of directions and indicators for assessing the different types of exclusions in the basic and advanced dimensions. A proposal to dimensionalize this assessment is presented in Table 6.2.

Table 6.2 Examples of proposals to dimensionalize the assessment of exclusion risks

Type of exclusion	*Basic measurement*	*Advanced measurement*
Civilizational	Percentage of households with access to running water and sanitation Electrification rate Poverty rate, including energy poverty Infant mortality rate Percentage of population living in slums	None: due to the treatment of baseline measurement as crucial for further evaluation (if a city does not achieve baseline indicators at a civilization-acceptable level, it does not guarantee the basic quality of life for its residents and cannot be considered smart in any other dimension)
Economic	Disposable income Gini coefficient Housing conditions (availability of housing, number of m^2 per person) Labor market (wage level; unemployment level; employment of minors)	Real estate prices Level of rents Level of the city's deficit and debt
Public safety	Guaranteeing the protection of human rights Level of crime Level of corruption	Availability and response time of preventive services (police, municipal police, sanitary inspection)
Educational	Availability of primary and secondary schools (enrolment rates)	Availability and quality of higher education institutions (enrolment rates; university rankings, monitoring of graduates' careers)
Health	Number of hospital beds, doctors and nurses per thousand residents Availability of medicines and vaccinations	Availability of online medical services Availability of applications that monitor the health of residents
Transportation	Availability of roads and public transport (car and rail)	Availability of air transport (number of airports; distance from the airport; number of air trips) Established car sharing system Availability of public bicycles and bicycle paths (per capita)

Table 6.2 (Continued)

Type of exclusion	*Basic measurement*	*Advanced measurement*
		Use of low-carbon sources of public and private transportation Forms of subsidies for sustainable mobility Availability of applications to support the mobility of residents
Social	The scale of possible discrimination based on gender, age, sexual orientation, religion, race or disability in social life	The scale of possible discrimination based on gender, age, sexual orientation, religion, race or disability at work and in education
Environmental	Level of pollution of water, air, soil Percentage of households with access to waste disposal	Percentage of wastewater treatment and waste recycled Size and availability of green areas Monitoring of the state of the environment available online or in apps for residents
Cultural and sports	Number of cinemas, theatres, museums per capita Number of sports facilities per capita	Subsidized participation of residents in sports and cultural events (especially those at risk of exclusion due to age, disability or income) Availability of information about cultural and sports events online / in apps, with the ability to book and purchase tickets remotely
Digital	Number of computers and telephones per capita Internet availability and speed	Scope of e-services offered to residents Availability of the Internet in public spaces Scope and quality of teaching IT competencies in primary, secondary and higher schools

Source: Authors' own work.

The concept of assessing the inclusiveness of Smart Cities presented above includes general dimensions and an assessment scheme. An issue that requires individual refinement is the choice of measures for assessing the risk of exclusion. It is also important to determine the levels of acceptability of indicators, especially those in the basic group determining the sufficiency of civilizational, economic, social or environmental conditions. In this regard, the authors suggest using generally accepted standards for minimum levels set by international organizations (zero-one system: meets the minimum or not) or referring to the leaders of the rankings of Smart Cities (threshold system, e.g., constitutes at least 30% of the value of the leader indicator). With this approach, it is important to be consistent in the use over time and throughout the adopted detailed assessment methodology.

Bibliography

Aliyev, A. G. (2021). Methodological basis of the comparative evaluation of inclusiveness level of economic development. *Management Dynamics in the Knowledge Economy*, 9(4), 404–418.

Bleja, J., Langer, H., Grossmann, U., Mörz, E. (2020). Smart cities for everyone: Age and gender as potential exclusion factors. *IEEE European Technology and Engineering Management Summit*, 5–7 March 2020, Dortmund, 1–5. https://doi.org/10.1109/E-TEMS46250.2020.9111741.

Bridgman, T., Cummings, S., Ballard, J. (2018). Who built Maslow's pyramid? A history of the creation of management studies' most famous symbol and its implications for management education. *Academy of Management Learning & Education*, 18(1), 81–98. https://doi.org/10.5465/amle.2017.0351.

Caird, S. P., Hallett, S. H. (2018). Towards evaluation design for smart city development. *Journal of Urban Design*, 24(2), 188–209. https://doi.org/10.1080/13574809.2018.1469402.

Dalprà, M. (2020). Inclusive playgrounds: A reality or a utopia in our cities? *Journal of Civil Engineering and Architecture*, 14, 635–643. https://doi.org/10.17265/1934-7359/2020.12.001.

de Castro Neto, M., de Melo Cartaxo, T. (2021). Algorithmic cities: A dystopic or utopic future? In: Aldinhas Ferreira, M. I. (ed), *How smart is your city? Intelligent systems, control and automation.* Science and Engineering, vol. 98. Cham: Springer. https://doi.org/10.1007/978-3-030-56926-6_6.

Doshi, S., Ranganathan, M. (2016). Contesting the unethical city: Land dispossession and corruption narratives in urban India. *Annals of the American Association of Geographers*, 107(1), 183–199. https://doi.org/10.1080/24694452.2016.1226124.

Gangadharan, S. (2021). Digital exclusion: A politics of refusal. In Bernholz, L., Landemore, H., Reich, R. (eds), *Digital technology and democratic theory*. Chicago: University of Chicago Press, 113–140. https://doi.org/10.7208/9780226748603-005.

Giddens, A. (2004). *Socjologia*. Wydawnictwo Naukowe PWN: Warszawa.

Haimes, Y. Y. (2015). *Risk modeling assessment and management*. Hoboken, NJ: Wiley.

Hassankhani, M., Alidadi, M., Sharifi, A., Azhdari, A. (2021). Smart city and crisis management: Lessons for the Covid-19 pandemic. *International Journal of Environmental Research and Public Health*, 18, 7736. https://doi.org/10.3390/ijerph18157736.

Holcombe, R. G. (1997). A theory of the theory of public goods. *Review of Austrian Economics*, 10(1), 1–22.

Kempin Reuter, T. (2019). Human rights and the city: Including marginalized communities in urban development and smart cities. *Journal of Human Rights*, 18(4), 382–402. https://doi.org/10.1080/14754835.2019.1629887.

Kim, K. (2023). Inclusion, exclusion, and participation in digital polis: Double-edged development of poor urban communities in alternative smart city-making. In Kim, K., Chung, H. (eds), *Gated communities and the digital polis*. Advances in 21st Century Human Settlements. Singapore: Springer. https://doi.org/10.1007/978-981-19-9685-6_9.

Knight, F. (1964). *Risk, uncertainty and profit*. New York: Hard, Schaffner & Marx.

Kolotouchkina, O., Barroso, C. L., Sanchez, J. L. M. (2022). Smart cities, the digital divide, and people with disabilities. *Cities*, 123, 103613. ttps://doi.org/10.1016/j.cities.2022.103613.

Maslow, A. H. (1943). Conflict, frustration, and the theory of threat. *Journal of Abnormal Psychology*, 38, 81–86.

Nizam, R., Karim, Z. A., Rahman, A. A., Sarmidi, T. (2020). Financial inclusiveness and economic growth: New evidence using a threshold regression analysis. *Economic Research-Ekonomska Istraživanja*, 33(1), 1465–1484. https://doi.org/10.1080/1331677X.2020.1748508.

Pereira, G. V., Cunha, M. A., Lampoltshammer, T. J., Parycek, P., Testa, M. G. (2017). Increasing collaboration and participation in smart city governance: A cross-case analysis of smart city initiatives. *Information Technology for Development*, 2(3), 526–553. https://doi.org/10.1080/02681102.2017.1353946

Pięta-Kanurska, M. (2019). Smart city a rozwój inkluzywny. *Biuletyn Komitet Przestrzennego Zagospodarowania Kraju Polskiej Akademii Nauk*, 273, 59–70.

Pritchard, C. L. (2014). *Risk management: Concepts and guidance*. Boca Raton: Taylor & Francis.

Reimer, B. (2004). Social exclusion in a comparative context. *Sociologia Ruralis*, 44(1), 75–78.

Richardson, L., Le Grand J. (2002). Outsider and insider expertise: The response of residents of deprived neighbourhoods to an academic definition of social exclusion. *CASE Papers*, 57, 10–11.

Rosati, U., Conti, S. (2016). What is a smart city project? An urban model or a corporate business plan? *Procedia – Social and Behavioral Sciences*, 223, 968–973.

Samuelson, W. F., Marks, S. G. (2009). *Ekonomia menedżerska.* Polskie Wydawnictwo Ekonomiczne: Warszawa.

Scott, K. (2016). Smart city Seattle and geographies of exclusion. In *The digital city and mediated urban ecologies*. Cham: Palgrave Macmillan, 45–61. https://doi.org/10.1007/978-3-319-39173-1_5.

Sharifi, A., Khavarian-Garmsir, A. R., Kummitha, R. K. R. (2021). Contributions of smart city solutions and technologies to resilience against the Covid-19 pandemic: A literature review. *Sustainability*, 13, 8018. https://doi.org/10.3390/su13148018.

Silver, H. (1994). Social exclusion and social solidarity: Three paradigms. *International Labour Review*, 133, 531–578.

Szopa, B., Szopa, A. (2011). Wykluczenie finansowe a wykluczenie społeczne. *Zeszyty Naukowe Polskiego Towarzystwa Ekonomicznego*, 11, 13–27.

Terje, A. (2015). Quantitative risk assessment. Cambridge: Cambridge University Press. https://doi.org/10.1017/CBO9780511974120

van Gils, B. A. M., Bailey, A. (2023) Revisiting inclusion in smart cities: Infrastructural hybridization and the institutionalization of citizen participation in Bengaluru's peripheries. *International Journal of Urban Sciences*, 27:sup1, 29–49. https://doi.org/10.1080/12265934.2021.1938640.

Wawer, M., Grzesiuk, K., Jegorow, D. (2022). Smart mobility in a smart city in the context of generation Z sustainability, use of ICT, and participation. *Energies*, 15, 4651. https://doi.org/10.3390/en15134651.

Wielicka-Gańczarczyk, K., Jonek-Kowalska, I. (2023). Involvement of local authorities in the protection of residents' health in the light of the smart city concept on the example of Polish cities. *Smart Cities*, 6, 744–763. https://doi.org/10.3390/smartcities6020036.

Willis, K.S. (2019). Whose right to the smart city? In Cardullo, P., Di Feliciantonio, C., Kitchin, R. (eds), *The right to the smart city*. Leeds: Emerald, 27–41. https://doi.org/10.1108/978-1-78769-139-120191002.

Yang, S., Chong, Z. (2021). Smart city projects against Covid-19: Quantitative evidence from China. *Sustainable Cities and Society*, 70, 102897. https://doi.org/10.1016/j.scs.2021.102897.

Zawieska, J., Pieriegud, J. (2018). Smart city as a tool for sustainable mobility and transport decarbonization. *Transport Policy*, 63, 39–50. https://doi.org/10.1016/j.tranpol.2017.11.004.

7 Long-term strategy for strengthening inclusiveness in the Smart City

7.1 Identification of future trends in exclusion in the Smart City

The discussion so far has identified many different types and forms of exclusion that can occur in a Smart City. However, it would be noteworthy to take a look at those that appear to be the most dangerous, and which are not necessarily given much attention in the literature. Admittedly, they are related to global and economy-wide trends, but their effects will certainly spill over to city residents, and it is the local authorities that will have to deal directly with social discontent.

According to the authors of this monograph, it is currently possible to identify 3 phenomena that can negatively affect the escalation of exclusion threats, both in Smart Cities and all other cities aspiring to be smart or wanting to improve the quality of life of urban communities. These are:

- Aging populations combined with declining female fertility rates (Mastnak et al., 2022; Kunkel and Settersten, 2021);
- Depletion of natural energy resources (Ahmad et al., 2023; Ali et al., 2021) combined with low use of renewable energy sources in emerging and developing economies (Zeren and Akkuş, 2020);
- Increasing population migrations of a known income nature, but also of a climate-related nature (Kalemba et al., 2022; Cummings et al., 2015).

In the remainder of this section, the above threats are documented using statistical data and references to the literature. Their links to urban exclusion are also identified.

DOI: 10.4324/9781003499992-8

The Smart City concept in its idealistic version implies the creation of strongly digitalized urban structures (Adedeji et al., 2022; Lukić et al., 2022; Marcu et al., 2020). The graphic illustration of such a Smart City appearing in many scientific and media publications depicts primarily young, healthy people and families with children. Such a vision ignores a significant portion of the actual urban community, which is made up of seniors, that is, people over 65 years. Given the demographic trends currently being observed, this is a very significant oversight.

In the literature, it is also very difficult to find publications treating the adaptation of the Smart City to the needs of the elderly, although there are articles treating children and young people (Pellegrino et al., 2022; Kuthar et al., 2021; van der Graaf, 2020). Exclusion due to age is, therefore, of little interest to researchers, despite the fact that with age cognitive abilities decline, problems with adaptation to new technologies appear and health problems increase, which can generate exclusion: digital, educational, health-related and exacerbate the scale of economic exclusion.

The growing scale of the threat of the problem of population aging is illustrated in Figures 7.1 and 7.2, which cover life expectancy and female fertility rates worldwide and in selected regions of the world, respectively. Also, Tables 7.1 and 7.2 show the average annual rate of change of the analysed parameters, as well as their total change in the period 1960–2022 and the arithmetic mean.

According to the Figure 7.1 data, since the beginning of the analysis, life expectancy has been steadily increasing in all regions studied. This is confirmed by the highly fitted linear increasing functions. A decrease in the growth rate is noticeable only in 2019–2021, which is related to the Covid-19 pandemic (Woolf, 2021).

By far, people who live in the Eurozone live the longest, which is due to the high level of development and economic prosperity in most European countries (Welsh et al., 2021; Weber and Loichinger, 2022; Tavares, 2022). The average life expectancy in the region exceeds 80 years. Nevertheless, regardless of the region, we are living longer and longer, which means that the number of seniors in urban communities around the world will increase steadily now and in the future.

It is also noteworthy that the change in life expectancy over the past 65 years is impressive, amounting to more than 41% worldwide, which is about 0.5% growth in this parameter per year. At the same time, life expectancy is also significantly increasing in less

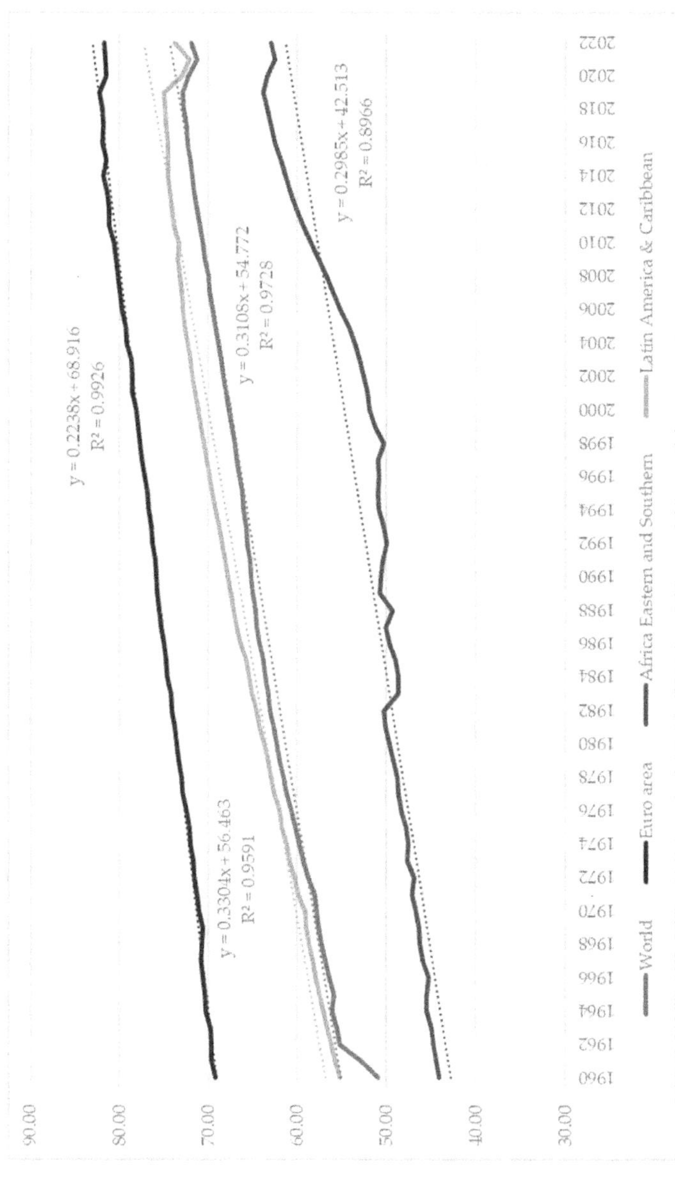

Figure 7.1 Life expectancy from 1960 to 2022 (in years).

Source: Authors' own compilation based on the World Bank data.

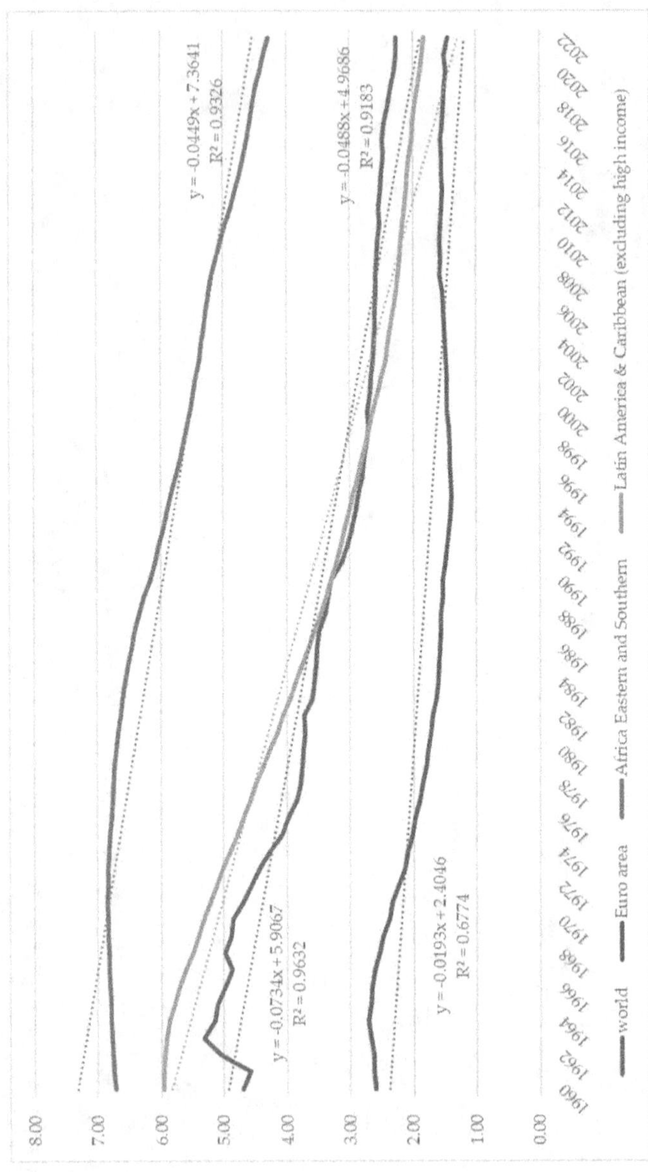

Figure 7.2 Female fertility in 1960–2022 (in number of births per woman).

Source: Authors' own compilation based on the World Bank data.

Table 7.1 Arithmetic mean, total change and average annual rate of change in life expectancy from 1960 to 2022

Parameter	*Arithmetic mean*	*Total change*	*Average annual rate of change*
World	64.72	41.46%	0.56%
Euro area	76.08	18.02%	0.27%
Africa Eastern and Southern	52.06	42.67%	0.57%
Latin America & Caribbean	67.03	33.72%	0.47%

Source: Authors' own study.

Table 7.2 Arithmetic mean, total change and average annual rate of change in female fertility from 1960 to 2022

Parameter	*Arithmetic mean*	*Total change*	*Average annual rate of change*
World	3.41	–27.46%	–0.52%
Euro area	1.79	–31.25%	–0.60%
Africa Eastern and Southern	5.93	–11.85%	–0.20%
Latin America & Caribbean	3.56	–40.27%	–0.83%

Source: Authors' own study.

economically developed regions (Africa, Latin America) (Ibrahim, 2022), which, on the one hand, shows the improvement of living conditions in these areas, but, on the other hand, also indicates the growing threat of population aging.

This threat is also intensified by the systematically and rapidly declining female fertility rate (Skakkebæk et al., 2022; Smulyanskaya, 2020), as shown in Figure 7.2. Additional data on this subject are presented in Table 7.2.

According to Figure 7.2, a systematic and significant downward trend (very well matched) characterizes fertility rates worldwide (Aitken, 2022), including in East and Southern Africa (Pourreza et al., 2021), as well as in Latin America and the Caribbean. Over 65 years, birth rates have declined by more than 27% on average worldwide, and in some regions by more than 40%. It has also fallen significantly in the Eurozone with the lowest baseline, illustrating the accumulation of demographic problems in that region (a fertility rate of below 2.10–2.15 does not guarantee replacement of generations).

Increasing life expectancy and declining female fertility negatively affect the age of the urban community, which, in practice, means an increase in the percentage of seniors and an increase in the risk of their exclusion due to problems associated with old age. However, it seems that modern cities downplay or fail to recognize such a threat, as they do not extensively analyse the needs of the senior part of the community. Nor do they always adapt urban infrastructure to the problems of reduced mobility, cognition or poorer health. Meanwhile, the aging of urban populations generates new needs and the need to modify existing infrastructure solutions. Thus, the authors note that there is a significant research gap in the area in question that requires deeper research and analysis, because, in light of the research presented here, the demographic crisis may significantly change the face of Smart Cities, especially those in Europe (and thus a significant portion of those existing and rated highest).

The image of Smart Cities has already been significantly altered by the migration crisis, particularly intense in Europe, due to the influx of refugees from Africa (Uchehara, 2016; Whitaker, 2017) and Ukraine. This has led to an increase in interest and publications from the area of migrant exclusion, which can relate not only to country of origin or cultural difference, but also to race or religion (Hassan 2020; De Genova, 2016; Bahri, 2021), factors that strongly polarize public sentiment.

Figure 7.3 shows the number of migrants in the EU and selected EU countries from 2014 to 2023. The data presented there shows that, on average, more than 760,000 people from African countries migrated to the EU annually, a number equivalent to the population of 1 large city. The record years in terms of migration were: 2015–2016 and 2023. The cited data also shows that the number of asylum seekers has been steadily increasing since 2020. This trend is also replicated in the European countries receiving the largest number of refugees, which include Germany, France and Italy.

Thus, according to Figure 7.4, in 2023, Germany received more than 30% of total migrants. France and Italy around 12–13%. Refugees also found a new place to live in Austria, the Netherlands and Switzerland.

The main reason for the migration of people from Africa to Europe is the desire for improved livelihoods and economic conditions (Fengler et al., 2020; Idemudia and Boehnke, 2020). This is clearly documented by the main migration destinations, which are highly

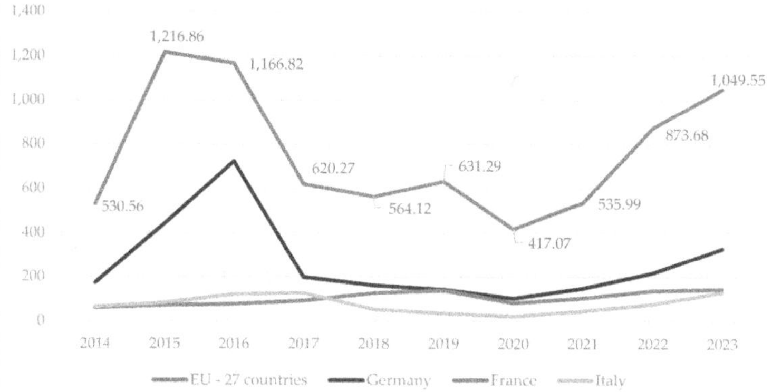

Figure 7.3 Number of migrants in EU and selected EU Member States in 2014–2023 (in thousands).

Source: Authors' own study based on Eurostat data.

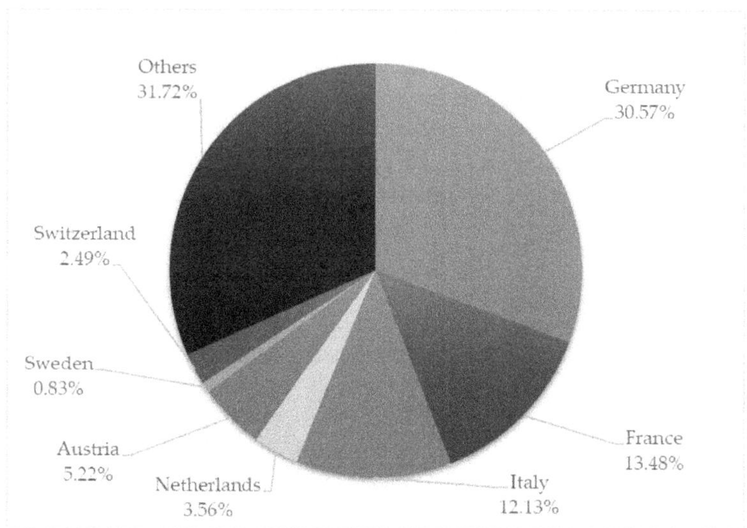

Figure 7.4 Structure of migrant admissions in selected EU Member States in 2023 (in percentage).

Source: Authors' own study based on Eurostat data.

developed European economies. It is also notable that these are also the countries with the largest number of cities recognized in prestigious rankings as Smart Cities. Migrations will, therefore, have a significant impact on the demographics of these cities. They can also become a source of serious social conflict and exclusion, which is already noticeable in all those countries that receive the largest number of refugees. This exclusion is most often associated with racial affiliation, religion and cultural difference. Nonetheless, it often also relates to economic issues, as migrants are sometimes paid less. They also often do not take up employment, adding to the ranks of the unemployed living on convenient social benefits.

In the past 3 years, Europe has also been hit by a migration crisis related to the Russian-Ukrainian war (Yaroshenko et al., 2023; Maidanik, 2023). Germany, Poland and the Czech Republic received the largest number of refugees from the area (Duszczyk and Kaczmarczyk, 2022). In this case, the motive for migration was the immediate threat to health and life and, in the long run, the desire to maintain the current property and income situation.

Mention should also be made of the coming wave of climate migrations, which are and will be associated with unfavourable climate change that will make it impossible for the population living in the threatened areas to continue. They include Sub-Saharan Africa, South Asia and Latin America. It is estimated that the population forced to resettle by 2050 will be around 143 million (www.gov.pl/web/europejska-siec-migracyjna/prognozy-wskazuja-ze-do-roku-2050-pojawi-sie-143-mln-migrantow-klimatycznych). This will place an additional burden on modern cities, which should already be thought about and acted upon to prevent migrant exclusion.

Periodic or continuous energy crises may also be a significant threat to cities (Min et al., 2024; Farghali et al., 2023). Non-renewable resources are limited and systematically depleted (Huo and Peng, 2023; Mittal and Gupta, 2015). Meanwhile, modern urban structures are characterized by high demand for electricity. This can increase the risk of energy poverty associated with intermittent or inadequate energy supply or complete lack of access to electricity.

Current data on the described threat is shown in Figure 7.5. It shows that access to electricity in all analysed regions is steadily increasing over time. However, it is only in the Eurozone, where it reaches 100%. The global average and indications for the other selected regions are far from meeting the 7th Sustainable Development Goal (Elavarasan

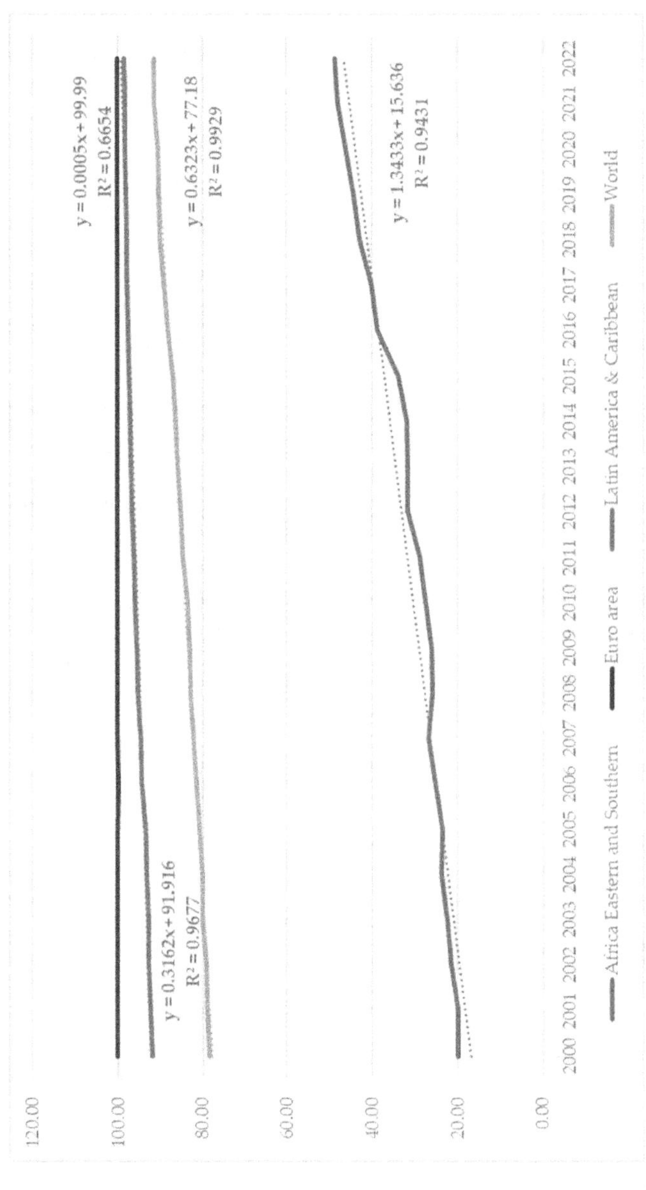

Figure 7.5 Access to electricity in 2000–2022 worldwide and in selected world regions (in percentage).

Source: Authors' own compilation based on the World Bank data.

et al., 2022; Dalei et al., 2021), which is to ensure affordable access to sources of stable, sustainable and modern energy for all.

The situation is worst in Africa, and the world average allows us to conclude (Ngarava et al., 2022; Ye and Koch, 2021) that about 9–10% of the population is still affected by energy poverty. The International Energy Agency also reports that 2023 was the first time in many years that electricity availability declined (www.iea.org/reports/sdg7-data-and-projections/access-to-electricity). With the reduction of non-renewable resources and the lack of progress in replacing them with zero-carbon energy sources, the risk of energy poverty may continue to increase (Hussain et al., 2023) due to rising energy prices (Goldthau and Tagliapietra, 2022). Thus, this is an important determinant of inclusiveness for Smart Cities, especially in areas with lower levels of civilizational and economic development. In this context, the literature exposes the link between energy exclusion and exclusion by race (Lin and Okyere, 2022).

The risks of massive socio-economic trends characterized above will certainly affect the inclusiveness of Smart Cities, as is already evident in the case of the migration crisis. For these reasons, it is important to have proper urban management, including the preparation of strategies to strengthen resilience to exclusion in the Smart City well in advance. After all, ad hoc action is never as effective as an anticipatory strategic approach. With the above in mind, the next section outlines key assumptions for a strategy to strengthen exclusion resilience in the Smart City.

7.2 Strategy to strengthen resistance to exclusions in the Smart City

A strategy is a long-term action plan aimed at achieving the stated strategic goal and operational objectives. For the inclusiveness strategy, the strategic goal is to strengthen the city's resilience to all kinds of exclusions. This is a very ambitious and difficult goal to achieve, due to the fact that it addresses many of the attitudes, opinions and prejudices that characterize the local community. Nonetheless, the variation in the level of inclusiveness, highlighted in the Smart City rankings, among others, shows that the drive to improve inclusiveness is yielding desirable results, although it is certainly sometimes far from perfection.

We should make it clear at this point that the creator and implementer of the strategy for reducing the risk of exclusion should be the municipal authorities as the coordinating body for the operation of urban infrastructure. However, this strategy should be created with the participation of all urban stakeholders, which, according to the fivefold helix, are (Paskaleva et al., 2021; Borghys et al., 2020; Susic et al., 2020; Dameri et al., 2016): (1) municipal authorities; (2) representatives of science and (3) business; (4) the urban community; and (5) representatives of environmental organizations, working to protect the urban environment. Such cooperation not only allows a holistic view of the strategy's assumptions, but in itself serves urban integration, which is one of the measures to prevent exclusion.

Thus, the starting point for developing a strategy to strengthen the city's resilience to exclusions should be an assessment of the risks associated with this social phenomenon. The concept of conducting it is presented in Chapter 6 of this monograph. The idea is to obtain answers to 2 questions: *What exclusions are present in the city and which may emerge in the future?* and *Does the scale of the identified exclusion threats require intervention and countermeasures?*

The checklist of exclusions provided in Table 6.1 will certainly be helpful to identify the threats of individual exclusions. In turn, an assessment of the scale of their occurrence can be carried out using the methodology described more extensively in Chapter 6. The results of this step will be the starting point for the selection of risks into those requiring intervention (countermeasures) or those requiring only monitoring and/or control. This selection should be carried out taking into account the exclusionary threat matrix shown in Figure 7.6.

The possible frequency of a given phenomenon (e.g. incidents of racial violence) can be assessed on the basis of statistical data, and the possible intensity can correspond to the feelings of the residents surveyed. In this way, the 2 dimensions above can be assigned to individual threats of exclusion. This, in turn, makes it possible to enter a threat into one of the quadrants shown in Figure 7.6.

For a low frequency of occurrence of a given threat and low intensity (Quadrant I), it is advisable to monitor a given exclusion to see if it becomes more serious in its consequences. Monitoring in this case means the systematic and continuous acquisition and analysis of information about a given hazard. As part of this monitoring, it is also necessary to check whether prevention of a given type of exclusion is

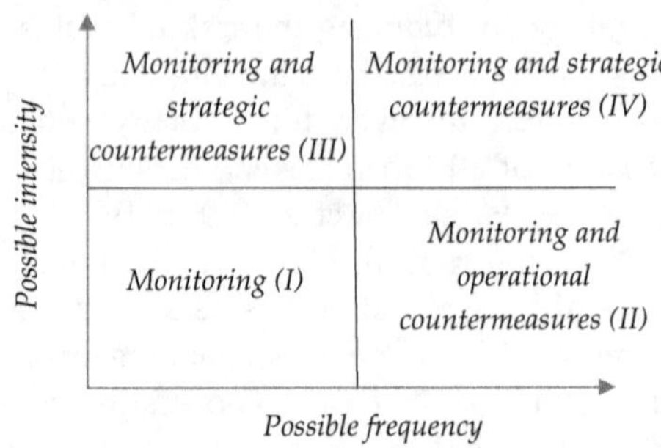

Figure 7.6 Exclusion threat assessment matrix.

Source: Authors' own work based on Staniec and Klimczak (2008) and Mauermann and Oktem (2002).

built into the city's usual activities, thus preventing its possible escalation (e.g. patrolling neighbourhoods inhabited by foreigners).

Of course, one can idealistically assume that the city tries to mitigate all, even the smaller threats of exclusion, nevertheless, in practice, due to the limited resources and means, this will not be possible. Hence the selection of threats and concentration on those that pose a serious threat to the quality of life of residents.

Threats, whose frequency is high, but whose intensity and thus possible losses and social harm are low (Quadrant II, such as low accessibility of cultural institutions), should be monitored and operational countermeasures taken. These are certainly vexing risks, but they do not pose a significant threat to the deterioration of the quality of urban life (Ligarski and Owczarek, 2023; Ligarski and Wolny, 2021). Generalization of their nature is also possible, to conclude that most of them relate to exclusion related to the satisfaction of higher-order needs. Operational activities in this case may involve organizing the city's ongoing support for residents as organizational or financial resources become available (e.g. subsidizing access to theatres or museums, organizing open days at cultural institutions).

Undoubtedly, threats, whose frequency is low (Quadrant III) or high (Quadrant IV), while intensity is high, which translates into the possibility of significant damage and significant deterioration in the quality of life of residents (e.g. the influx of a significant number of migrants to the city), should be considered the most dangerous. Such

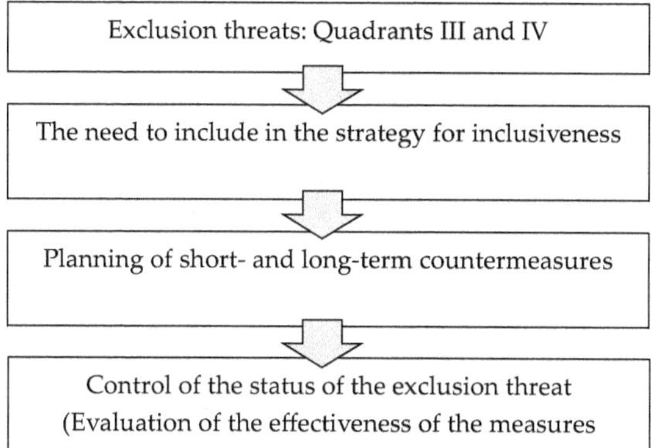

Figure 7.7 Scheme of action for identifying the risks of exclusion from Quadrants III and IV of the matrix.

Source: Authors' own study.

threats need to be written into the strategy for increasing Smart City inclusiveness and take systematic and long-term countermeasures of a strategic nature, according to the diagram shown in Figure 7.7.

The following section focuses on general principles for determining countermeasures in an inclusiveness strategy. Nevertheless, it is worth emphasizing that the diversity of identified exclusions does not allow the development of a single closed or universal catalogue of conduct or behaviour. For this reason, further considerations are synthetic and, in some cases, exemplary.

Thus, the identified types of exclusions can be divided into 2 more general groups. The first is related to urban infrastructure. The second relates to the urban community, and thus has a humanistic dimension. Such a division alludes to the duality of the Smart City, which was described more extensively in the initial chapters of this monograph. With this in mind, countermeasures in the 2 groups of exclusions will differ. This is because the quality and accessibility of urban infrastructure is affected differently. Relationships between residents are affected differently. Although it is worth adding that the exclusions analysed are very often interdependent and intermingle with each other, which is characteristic of the multidimensionality and diversity of urban life.

Countermeasures of an infrastructural nature refer primarily to the threats of exclusion: civilizational, economic, public safety, educational, health, communication, environmental, cultural and sports, as well as digital, that is, most of the types catalogued earlier. This is because, in practice, these threats mean the absence, limited availability or poor quality of specified urban infrastructure. The city must, therefore, in relation to the above shortfalls, determine: the cause and extent of limited availability of urban infrastructure. Unfortunately, most often the cause of limited availability is a lack of financial resources, indicating, again, the strong dependence of exclusion threats on the economic situation of the city and the region, in which it is located. Then the only countermeasure becomes raising additional funds to supplement, expand, create new infrastructure in the city. City budgets everywhere are limited, but alternative forms of financing can be sought, such as public–private partnerships, crowd-funding or national or international infrastructure projects.

A slightly different cause of exclusion of an infrastructural nature is the limitation of infrastructure accessibility for certain social groups, such as seniors, people with disabilities or migrants. Then, the attention of countermeasure decision-makers should focus, first, on identifying barriers that limit accessibility, which can be financial, technical or mental.

Each of these requires different countermeasures. Income barriers can be addressed by discounts or the elimination of fees for the use of a given infrastructure (such as theatres or public transportation). Technical barriers should be eliminated at the infrastructure design stage and, if this has not happened, excluded groups should be offered facilities *ex post* (e.g. an elevator for people with mobility disabilities at the design stage or a stairlift as an amenity). Mental barriers (e.g. fear of urban applications or inability of seniors to use them) should be reduced through education, promotion and integration of residents.

The concept of taking countermeasures for infrastructure exclusions is summarized in Figure 7.8.

Exclusions of a human, social nature are quite different and require a different approach in terms of mitigation. They are often an offshoot of the development of Smart Cities and systematic urbanization, which involve attracting more and more new residents, lured by the search for a better quality of life. Combating them also seems to be more challenging, as social exclusions are associated with deeply held beliefs, customs and stereotypes. They are also represented by a

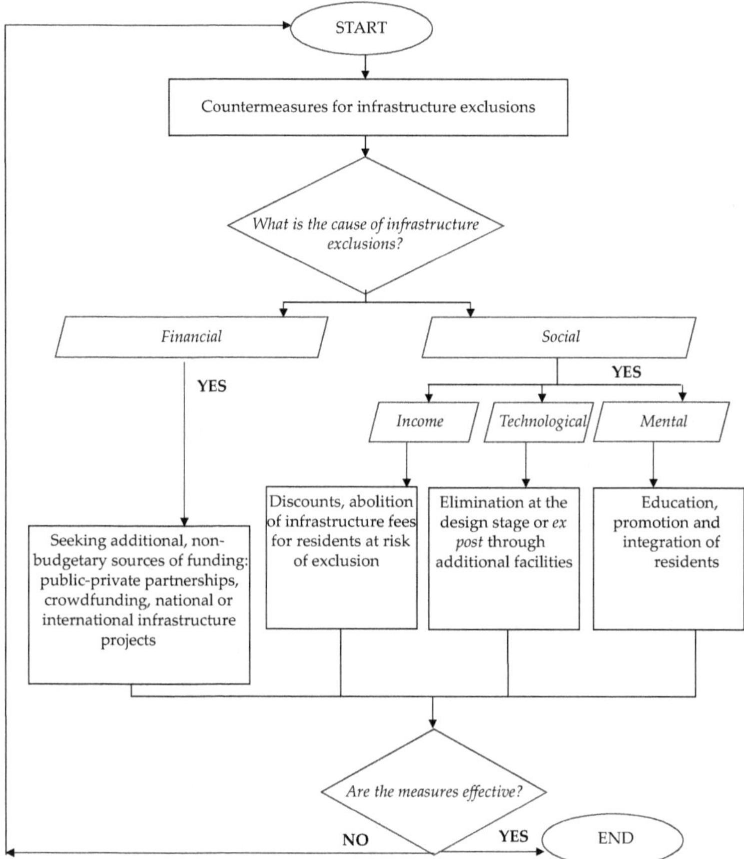

Figure 7.8 Concept of countermeasures in case of infrastructure threats.

Source: Authors' own study.

wide variety of individuals and social groups, which may necessitate a strong diversity of countermeasures.

The approach proposed in business as part of diversity management can be useful in reducing social exclusions (Sitko-Lutek and Bednarzewska, 2016; Rakowska, 2016; Rakowska and Sitko-Lutek, 2016; Rakowska, 2023). It is one of the human resource management concepts used in today's highly internationalized companies, which also face the challenge of otherness. At the core of diversity management is the fight against discrimination at work based on race, religion or gender, age or sexual orientation. The reasons for exclusions are, therefore, the same, but the environment, in which they operate is different. Nevertheless, some of the diversity management measures are suitable for adaptation to the needs of city governments for strengthening Smart City inclusiveness. These are described below and expanded upon with the authors' own observations in this monograph.

Diversity is and will continue to be a permanent feature of an urban community. It implies the existence of visible and invisible differences between individuals (residents). It can include attitudes, behaviour, social or material status, gender, age, religion and many other characteristics (Sitko-Lutek, Ławicka-Kruk and Jakubiak, 2024; Rakowska and Cichorzewska, 2016, 2019). In the variant of full inclusiveness, diversity does not generate tensions and conflicts and does not constitute grounds for discrimination. Such an ideal state, however, does not exist in practice, making diversity management a necessity to protect modern cities from the negative effects of exclusion.

Walczak (2011a, 2011b) distinguishes 3 dimensions of diversity, 2 of which can be applied to urban communities. They are related to a person's primary and secondary identity. Primary identity stems from within the individual and is related to their psychophysical characteristics. It is also derived from their education, knowledge, value system, upbringing, belief norms and personal culture. Primary diversity is, therefore, related to: gender, age, ethnicity, religion, professed values, as well as competence, aptitude, personality traits.

Secondary identity is acquired and results from a person's place in the community. It is derived from the roles and functions that a person performs at work and in private life. The diversity that results from this identity usually concerns: material status, social status, place of residence, appearance, education, experience, having children, interests, etc. (Walczak, 2011a, 2011b).

Regardless of its origin, diversity must be managed (Skrzypek, 2018a, 2018b) (responsibility for this task, as already mentioned, belongs to the city government), which, in practice, means levelling the playing field for all residents to have a better life in Smart Cities. Governance understood in this way can be looked at from a narrow perspective, which will be limited only to the prevention of exclusion and related discrimination through prevention and penalization of undesirable behaviour.

It can also be implemented in a much broader context, which means taking 3 types of actions: (1) the elimination of undesirable behaviour; (2) the integration of the city's diverse social groups to ensure equal opportunities for all, and thus a sense of inclusiveness; and (3) the use of diversity as a source of strategic advantage for the city, a distinguishing feature that testifies to the pursuit of fully sustainable development of the urban community (Guillaume et al., 2017; Urbaniak, 2014). In the context described above, the listed stages may also reflect the city's maturity in terms of inclusiveness.

The latter stage may seem idealistic and unrealistic in today's socio-economic conditions. Nevertheless, it should be remembered that certain phenomena, events and processes (e.g. migration) can no longer be stopped and, therefore, it is necessary to create such conditions for urban life as soon as possible, so that diversity is not a problem and does not contribute to the deterioration of the quality of urban existence.

Interestingly, diversity can be a source of benefits (Baggio et al., 2022; Ely and Thomsa, 2020; Gurin et al., 2004; Smith and Schonfeld, 2000), among which we can point out the following:

- Exchanging knowledge and experience that fosters entrepreneurship and innovation, as well as strengthens the city's competitiveness;
- Having residents, who are complementary in terms of skills and competencies, stimulating the development of the labour market;
- Opening up to new needs and expectations from the city's infrastructure;
- Improving communication and public services oriented to meet new expectations;
- Strengthening empathy and openness to the other in a way that significantly humanizes modern cities and makes them more sustainable;
- Improving the image of the city.

The key to success in diversity management, and thus in the pursuit of inclusiveness, is undoubtedly a change in mindset (Walczak, 2011a, 2011b), which must occur, first and foremost, on the part of the city authorities. The way of thinking and approach to diversity must also be changed by other urban stakeholders, that is, residents,

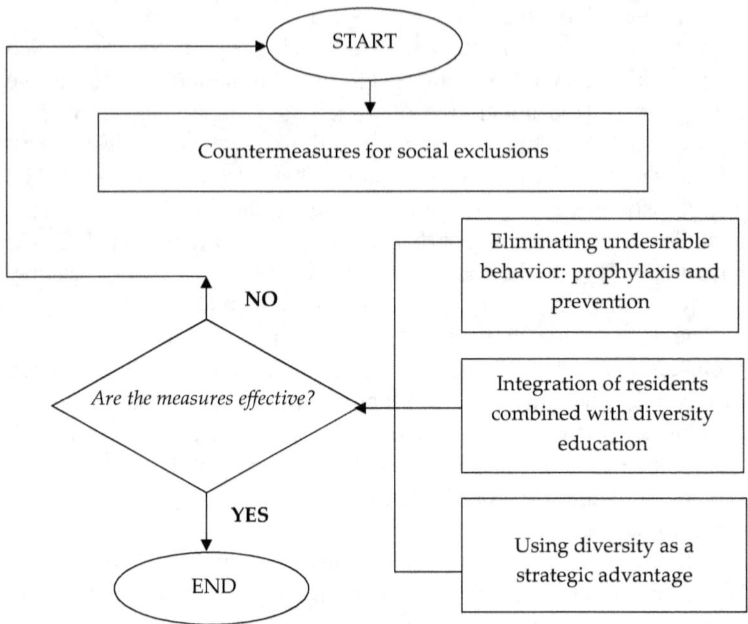

Figure 7.9 Concept of countermeasures in case of social risks.

Source: Authors' own study.

representatives of business and science and other organizations operating in the city.

The concept of addressing social exclusions is summarized in Figure 7.9.

Besides identifying and assessing the risks of urban exclusion, as well as developing strategies to strengthen urban inclusiveness, it is important to monitor the status of the exclusion threat, and thus assess the effectiveness of the measures taken in this regard. It can be carried out periodically (e.g. annually) by comparing the scale of the exclusion risk, taking into account objective assessment (statistical data) and subjective assessment (surveys) before and after the implementation of specific measures. Then, the comparison obtained will allow to assess the degree of implementation of the intentions for inclusiveness, thus the effectiveness of the implementation of individual strategic objectives.

Bibliography

Adedeji, K. B., Ponnle, A. A., Abu-Mahfouz, A. M., Kurien, A. M. (2022). Towards digitalization of water supply systems for sustainable smart city development: Water 4.0. *Applied Sciences*, 12, 9174. https://doi.org/10.3390/app12189174.

Ahmad, M., Ahmed, Z., Yang, X., Can, Y. (2023). Natural resources depletion, financial risk, and human well-being: What is the role of green innovation and economic globalization? *Social Indicators Research*, 167, 269–288. https://doi.org/10.1007/s11205-023-03106-9.

Aitken, R. J. (2022). The changing tide of human fertility. *Human Reproduction*, 37(4), 629–638. https://doi.org/10.1093/humrep/deac011.

Ali, A., Audi, M., Roussel, Y. (2021). Natural resources depletion, renewable energy consumption and environmental degradation: A comparative analysis of developed and developing world. *International Journal of Energy Economics and Policy*, 11(3), 251–260. www.econjournals.com/index.php/ijeep/article/download/11008/5814.

Baggio, J. A., Freeman, J., Coyle, T. R., Anderies, J. M. (2022). Harnessing the benefits of diversity to address socio-environmental governance challenges. *PloS One*, 17(8), e0263399.

Bahri, M. T. (2021). Understanding the pattern of international migration: Challenges in human rights protection. *Jurnal Hukum Unissula*, 8(2), 81–98. https://ssrn.com/abstract=4261471.

Borghys, K., Van Der Graaf, S., Walravens, N., Van Compernolle, M. (2020). Multi-stakeholder innovation in smart city discourse: Quadruple helix thinking in the age of 'platforms'. *Frontiers in Sustainable Cities*, 2, 5.

Chewe, M., Hangoma, P. (2020). Drivers of health in sub-Saharan Africa: A dynamic panel analysis. *Health Policy OPEN*, 1, 100013. https://doi.org/10.1016/j.hpopen.2020.100013.

Cummings, C., Pacitto, J., Lauro, D., Foresti, M. (2015). *Why people move: Understanding the drivers and trends of migration to Europe*. London: Overseas Development Institute.

Dalei, N. N., Painuly, P. K., Rawat, A., Heggde, G. S. (2021). Sustainable energy challenges in realizing SDG 7. In Leal Filho, W., Azul, A. M., Brandli, L., Lange Salvia, A., Wall, T. (eds), *Affordable and clean energy*. Encyclopedia of the UN Sustainable Development Goals. Cham: Springer, 34–37. https://doi.org/10.1007/978-3-319-71057-0_157-1.

Dameri, R. P., Negre, E., Rosenthal-Sabroux, C. (2016). Triple Helix in smart cities: A literature review about the vision of public bodies, universities, and private companies. In *49th Hawaii International Conference on System Sciences*, 5–8 January 2016, Koloa, HI, USA 2974–2982.

De Genova, N. (2016). The European question: Migration, race, and postcoloniality in Europe. *Social Text*, 34, 75–102. https://doi.org/10.1215/01642472-3607588.

Duszczyk, M., Kaczmarczyk, P. (2022). The war in Ukraine and migration to Poland: Outlook and challenges. *Intereconomics*, 57, 164–170. https://doi.org/10.1007/s10272-022-1053-6.

Elavarasan, R. M., Pugazhendhi, R., Irfan, M., Mihet-Popa, L., Campana, P. E., Khan, I. A. (2022). A novel Sustainable Development Goal 7 composite index as the paradigm for energy sustainability assessment: A case study from Europe. *Applied Energy*, 307, 118173. https://doi.org/10.1016/j.apenergy.2021.118173.

Ely, R. J., Thomas, D. A. (2020). Getting serious about diversity. *Harvard Business Review*, 98(6), 114–122.

Farghali, M., Osman, A. I., Mohamed, I. M. A., Chen, Z., Chen, L., Ihara, I., Pow-Seng, Y., Rooney, D. W. (2023). Strategies to save energy in the context of the energy crisis: A review. *Environmental Chemistry Letters*, 21, 2003–2039. https://doi.org/10.1007/s10311-023-01591-5.

Fengler, S., Bastian, M., Brinkmann, J., Zappe, A. C., Tatah, V., Andindilile, M., Lengauer, M. (2020). Covering migration in Africa and Europe: Results from a comparative analysis of 11 countries. *Journalism Practice*, 16(1), 140–160. https://doi.org/10.1080/17512786.2020.1792333.

Goldthau, A., Tagliapietra, S. (2022). Energy crisis: Five questions that must be answered in 2023. *Nature*, 612, 627–630. https://doi.org/10.1038/d41586-022-04467-w.

Guillaume, Y. R., Dawson, J. F., Otaye-Ebede, L., Woods, S. A., West, M. A. (2017). Harnessing demographic differences in organizations: What moderates the effects of workplace diversity? *Journal of Organizational Behavior*, 38(2), 276–303.

Gurin, P., Nagda, B. R. A., Lopez, G. E. (2004). The benefits of diversity in education for democratic citizenship. *Journal of Social Issues*, 60(1), 17–34.

Hassan, H. A. (2020). Transformations of forced migration in Africa: Issues and general problems. *African Journal of Political Science and International Relations*. https://doi.org/10.5897/AJPSIR2020.1255.

Europejska Sieć Migracyjna: Prognozy wskazują, że do roku 2050 pojawi się 143 mln migrantów klimatycznych . Serwis Rzeczypospolitej Polskiej. www.gov.pl/web/europejska-siec-migracyjna/prognozy-wskazuja-ze-do-roku-2050-pojawi-sie-143-mln-migrantow-klimatycznych (accessed 10.08.2024).

Huo, J., Peng, Ch. (2023). Depletion of natural resources and environmental quality: Prospects of energy use, energy imports, and economic growth hindrances. *Resources Policy*, 86, part A, 104049. https://doi.org/10.1016/j.resourpol.2023.104049.

Hussain, S. A., Razin, F., Hewage, K., Sadiq, R. (2023). The perspective of energy poverty and 1st energy crisis of green transition. *Energy*, 275, 127487. https://doi.org/10.1016/j.energy.2023.127487.

Ibrahim, R. L. (2022). Beyond COP26: Can income level moderate fossil fuels, carbon emissions, and human capital for healthy life expectancy in

Africa? *Environmental Science and Pollution Research*, 29, 87568–87582. https://doi.org/10.1007/s11356-022-21872-w.

Idemudia, E., Boehnke, K. (2020). Patterns and current trends in African migration to Europe. In *Psychosocial experiences of African migrants in six European countries*. Social Indicators Research Series, vol. 81. Cham: Springer, 15–31, https://doi.org/10.1007/978-3-030-48347-0_2.

Kalemba, S. V., Bernard, A., Corcoran, J., Charles-Edwards, E. (2022). Has the decline in the intensity of internal migration been accompanied by changes in reasons for migration? *Journal of Population Research*, 39, 279–313. https://doi.org/10.1007/s12546-022-09285-5.

Khutar, D. Z., Yahya, O. H., ALRikabi, H. T. S. (2021) Design and implementation of a smart system for school children tracking. *IOP Conference Series: Materials Science and Engineering*, 1090, 012033. https://doi.org/10.1088/1757-899X/1090/1/012033.

Kunkel, S. R., Settersten, R. (2021). *Aging, society, and the life course*, 6th ed. New York: Springer.

Ligarski, M. J., Owczarek, T. (2023). How cities study quality of life and use this information: Results of an empirical study. *Sustainability*, 15, 8221. https://doi.org/10.3390/su15108221.

Ligarski, M. J., Wolny, M. (2021). Quality of life surveys as a method of obtaining data for sustainable city development: Results of empirical research. *Energies*, 14, 7592. https://doi.org/10.3390/en14227592.

Lin, B., Okyere, M. A. (2022). Race and energy poverty: The moderating role of subsidies in South Africa. *Energy Economics*, 117, 106464. https://doi.org/10.1016/j.eneco.2022.106464.

Lukić, I., Miličević, K., Köhler, M., Vinko, D. (2022). Possible blockchain solutions according to a smart city digitalization strategy. *Applied Sciences*, 12, 5552. https://doi.org/10.3390/app12115552.

Maidanik, I. (2023). The forced migration from Ukraine after the full scale Russian invasion: Dynamics and decision making drivers. *European Societies*, 26(2), 469–480. https://doi.org/10.1080/14616696.2023.2268150.

Marcu, I., Suciu, G., Bălăceanu, C., Vulpe, A., Drăgulinescu, A.-M. (2020). Arrowhead technology for digitalization and automation solution: Smart cities and smart agriculture. *Sensors*, 20, 1464. https://doi.org/10.3390/s20051464.

Mastnak, T., Maver, U., Finšgar, M. (2022). Addressing the needs of the rapidly aging society through the development of multifunctional bioactive coatings for orthopedic applications. *International Journal of Molecular Sciences*, 23, 2786. https://doi.org/10.3390/ijms23052786.

Mauermann, A., Oktem, U. (2002). The near-miss management in operational risk. *Journal of Risk Finance*, 4(1), 25–38.

Min, B., O'Keeffe, Z. P., Abidoye, B., Gaba, K. M., Monroe, T., Stewart, B. P., Baugh, K., Sánchez-Andrade Nuño, B. (2024). Lost in the dark: A survey of energy poverty from space. *Joule*, 8(7). https://doi.org/10.1016/j.joule.2024.05.001.

Mittal, I., Gupta, R. K. (2015). Natural resources depletion and economic growth in present era. *SOCH – Mastnath Journal of Science & Technology*, 10(3). https://ssrn.com/abstract=2920080.

Ngarava, S., Zhou, L., Ningi, T., Chari, M. M., Mdiya, L. (2022). Gender and ethnic disparities in energy poverty: The case of South Africa. *Energy Policy*, 161, 112755. https://doi.org/10.1016/j.enpol.2021.112755.

Paskaleva, K., Evans, J., Watson, K. (2021). Co-producing smart cities: A Quadruple Helix approach to assessment. *European Urban and Regional Studies*, 28(4), 395–412.

Pellegrino, M. A., Roumelioti, E., D'Angelo, M., Gennari, R. (2022). Children's participation in the design of smart solutions: A literature review. *Smart Cities*, 5, 475–495. https://doi.org/10.3390/smartcities5020026.

Pourreza, A., Sadeghi, A., Amini-Rarani, M., Khodayari-Zarnaq, R., Jafari, H. (2021). Contributing factors to the total fertility rate declining trend in the Middle East and North Africa: A systemic review. *Journal of Health, Population and Nutrition*, 40, 11. https://doi.org/10.1186/s41043-021-00239-w.

Rakowska, A. (2016). Potencjał kompetencyjny kadry kierowniczej innowacyjnych przedsiębiorstw w perspektywie zarządzania różnorodnością zasobów ludzkich – wyniki badań. *Zeszyty Naukowe Politechniki Śląskiej, seria: Organizacja i Zarządzanie*, 97, 239–252.

Rakowska, A. (2023). Kulturowa różnorodność pracowników w czasach nasilonej emigracji. *Zarządzanie Zasobami Ludzkimi*, 15(1). https://doi.org/10.5604/01.3001.0016.2922.

Rakowska, A., Cichorzewska, M. (2016). Zarządzanie różnorodnością zasobów ludzkich w innowacyjnych przedsiębiorstwach–wyniki badań. *Przedsiębiorczość i Zarządzanie*, 12(2–3), 89–100.

Rakowska, A., Cichorzewska, M. (2019). Rola wartości preferowanych przez pracowników w kontekście kształtowania inkluzywnych organizacji. *Przedsiębiorczość i Zarządzanie*, 20(6/3), 327–341.

Rakowska, A., Sitko-Lutek, A. (2016). Umiejętności i style uczenia się w zarządzaniu różnorodnością. *Zarządzanie i Finanse*, 14(2), 311–323.

Sitko-Lutek, A., Bednarzewska, K. (2016). Wykorzystanie znajomości stylów uczenia się w zarządzaniu różnorodnością. *Przedsiębiorczość i Zarządzanie*, 17, 149–158.

Sitko-Lutek, A., Ławicka-Kruk, K., Jakubiak, M. (2024). Diversity management in the medical device industry in the light of the results of international empirical research. *Krakow Review of Economics and Management/Zeszyty Naukowe Uniwersytetu Ekonomicznego w Krakowie*, 2(1004), 45–61.

Skakkebæk, N. E., Lindahl-Jacobsen, R., Levine, H., Andersson, A.-M., Jørgensen, N., Main, K. M., Lidegaard, O., Priskorn, L., Holmboe, S.A.,

Bräuner, E. V., Almstrup, K., Franca, L. R., Znaor, A., Kortenkamp, A., Hart, R. J., Juul, A. (2022). Environmental factors in declining human fertility. *Nature Reviews Endocrinology*, 18, 139–157. https://doi.org/10.1038/s41574-021-00598-8.

Skrzypek, E. (2018a). Zarządzanie różnorodnością-stan i perspektywy rozwoju. *Problemy Jakości*, 10, 4–11.

Skrzypek, E. (2018b). Nowoczesne trendy w zarządzaniu a doskonalenie zarządzania. https://open.icm.edu.pl/items/0547cc18-3b2b-45a4-a422-afeb78062c67 (accessed 17.08.2024).

Smith, D. G., Schonfeld, N. B. (2000). The benefits of diversity what the research tells us. *About Campus*, 5(5), 16–23.

Smulyanskaya, N. S. (2020). Factors of fertility ageing rate. *Population and Economics*, 4(1), 60–74. https://doi.org/10.3897/popecon.4.e53039.

Staniec, I., Klimczak, K. K. (2008). *Ryzyko operacyjne, [w:]* Zarządzanie ryzykiem operacyjnym, *pod red. I. Staniec i J. Zawiły-Niedźwieckiego.* Wydawnictwo C.H. Beck: Warszawa.

Suzic, B., Ulmer, A., Schumacher, J. (2020). Complementarities and synergies of quadruple helix innovation design in smart city development. In *2020 Smart City Symposium*, Prague, 25 June 2020, 1–7.

Syed, J., Özbilgin, M. (2009). A relational framework for international transfer of diversity management practices. *International Journal of Human Resource Management*, 20(12), 2435–2453.

Tavares, A. I. (2022). Life expectancy at 65, associated factors for women and men in Europe. *European Journal of Ageing*, 19, 1213–1227. https://doi.org/10.1007/s10433-022-00695-1.

Uchehara, K. (2016). Sub-Saharan African countries and migration to Europe: Exploring the motivations, effects and solutions. *Informatologia*, 49 (1–2), 79–85. https://hrcak.srce.hr/161903.

Urbaniak, B. (2014). Zarządzanie różnorodnością zasobów ludzkich w organizacji. *Zarządzanie Zasobami Ludzkimi*, 3–4, 98–99.

van der Graaf, S. (2020). The right to the city in the platform age: Child-friendly city and smart city premises in contention. *Information*, 11, 285. https://doi.org/10.3390/info11060285.

Walczak, W. (2011a). Zarządzanie różnorodnością wyzwaniem dla współczesnych organizacji. *Organization and Management*, 3(146), 43–58.

Walczak, W. (2011b). Zarządzanie różnorodnością jako podstawa budowania potencjału kapitału ludzkiego organizacji. *E-mentor*, 3(40), 11–19.

Weber, D., Loichinger, E. (2022). Live longer, retire later? Developments of healthy life expectancies and working life expectancies between age 50–59 and age 60–69 in Europe. *European Journal of Ageing*, 19, 75–93. https://doi.org/10.1007/s10433-020-00592-5.

Welsh, C. E., Matthews, F. E., Jagger, C. (2021). Trends in life expectancy and healthy life years at birth and age 65 in the UK, 2008–2016, and other countries of the EU28: An observational cross-sectional study. *Lancet. Regional Health*, 2, 100023. https://doi.org/10.1016/j.lanepe.2020.100023.

Whitaker, B. E. (2017). Migration within Africa and beyond. *African Studies Review*, 60(2), 209–220. https://doi.org/10.1017/asr.2017.49.

Woolf, S. H. (2021). Effect of the Covid-19 pandemic in 2020 on life expectancy across populations in the USA and other high income countries: simulations of provisional mortality data. *BMJ*, 373. https://doi.org/10.1136/bmj.n1343.

Yaroshenko, O. M., Smorodynskyi, V. S., Silchenko, S. O., Vetukhova, I. A., Zaika, D. I. (2023). The crisis of labor migration during the Russia-Ukraine War. *New Labor Forum*, 32(3), 80–88. https://doi.org/10.1177/10957960231195621.

Ye, Y., Koch, S. F. (2021). Measuring energy poverty in South Africa based on household required energy consumption. *Energy Economics*, 103, 105553. https://doi.org/10.1016/j.eneco.2021.105553.

Zeren, A. (2020). The relationship between renewable energy consumption and trade openness: New evidence from emerging economies. *Renewable Energy*, 147(1), 322–329. https://doi.org/10.1016/j.renene.2019.09.006.

8 Inclusiveness versus exclusivity

Discussion and conclusions

8.1 Discussion in the light of current research and practices

The findings presented in the monograph confirm that the development of technology in Smart Cities carries a significant risk of exacerbating social and digital exclusion, as has been previously pointed out by researchers working on this topic. For example, previous studies, such as those by Vanolo (2014) and Kitchin (2015), have pointed out that the development of technology can lead to further marginalization of social groups that, for various reasons, have limited access to modern technology. The monograph confirms these concerns, showing that Smart Cities can create a new kind of exclusion, especially for the elderly, people with lower levels of education and low-income residents, who may have difficulty using advanced digital services.

One of the key themes explored in both the monograph and earlier studies is the potential of ICT to transform cities into more efficient and sustainable environments. Notably, Smart City studies, such as those by Townsend (2013) and Kitchin (2016), have emphasized that technologies can help improve the quality of life for residents, but, at the same time, risk excluding social groups that have limited access to these technologies for various reasons. The monograph confirms these concerns, stressing that the development of technology, if not properly managed, can lead to further marginalization of the elderly, lower-income people and groups with less access to technological education. In this regard, the analytical results presented in the monograph are consistent with previous studies, but, at the same time, offer new insights into the mechanisms that can lead to such exclusion, especially in the context of the increasing digitization of urban services

DOI: 10.4324/9781003499992-9

and other socio-economic phenomena, such as financial crises, pandemics, migration and aging populations.

The monograph also presents positive examples where Smart City development contributes to reducing exclusion. The authors of the monograph emphasize the importance of inclusive design of urban technologies, which is confirmed by studies, such as those by Calvillo et al. (2016), who noted that the appropriate implementation of information and communication technologies can improve the accessibility of urban services for groups at risk of marginalization. However, it is necessary to identify, recognize and monitor the needs of the urban community from the bottom up, so as to skilfully combine modern technologies with the demand for inclusiveness.

The monograph also draws attention to the need to consider the local context in Smart City-related analyses, which is in line with the findings of Shelton and Lodato (2019), who stressed that Smart City policies must be tailored to the specific social, economic and cultural conditions of a given city. The monograph elaborates on this concept, pointing out that a lack of such customization can lead to a situation, in which only select social groups benefit from the development of technology, while others are increasingly marginalized. The authors suggest that an overly general or project-based approach to technology implementation in cities may not only exacerbate existing inequalities, but also create new forms of exclusion, the scale and nature of which may vary depending on local conditions. The above is particularly evident in cities, where there is a significant technological gap between different social groups. This issue was also explored by Van Deursen and Van Dijk (2014).

The monograph also presents a discussion of the impact of Smart Cities on access to public services. Previous studies, such as those by Hollands (2008), have criticized the Smart Cities concept for focusing mainly on technological efficiency at the expense of social equity. The authors of the monograph bring new elements to this discussion, pointing out that technology can both facilitate access to services and hinder it, depending on how it is implemented. For example, the implementation of advanced transportation management systems can improve urban mobility, but, at the same time, can exclude people, who do not have access to the right digital tools. This dual nature of technology in the context of Smart Cities has also been previously described by Kitchin (2016), but the threads described in the monograph bring to this discussion a detailed analysis of specific cases that illustrate how different social groups can perceive and be affected by technological solutions.

The monograph also emphasizes the importance of civic participation and transparency in decision-making processes regarding urban technology development. Previous research, such as that by Cardullo and Kitchin (2019), has already pointed to the importance of citizen involvement in city planning processes, but the monograph's authors expand on this concept, suggesting that participation should be at the heart of Smart City strategies to effectively address exclusion and build more sustainable urban ecosystems. The monograph expands on this concept, suggesting that participation should not just be an addition to decision-making processes, but an integral part of them. In doing so, the authors emphasize that a lack of citizen involvement can lead to the reinforcement of existing inequalities and exclusion, as decisions made without extensive consultation with residents often fail to take into account their real needs and expectations. Meanwhile, the relevant literature is decidedly lacking civic perspective. Considerations are mainly made from the perspective of the city, business and science, which can be a source of serious distortions of the Smart City concept.

The monograph also considers issues of data analysis and its role in city management. In the literature on the subject, authors such as Batty (2013) have emphasized that data is a key element in the development of Smart Cities, but the monograph takes this analysis a step further, pointing out the risks associated with inappropriate use of data, especially in the context of privacy and ethics. The authors argue that data can be both a tool for making cities more efficient and inclusive, and a source of new forms of exclusion if not used properly.

Recommendations for Smart City authorities to incorporate exclusion resilience into their policies should be based on a comprehensive approach that integrates social, technological and urban issues to create more inclusive and equitable urban environments. In an era of rapid technological development and increasing urbanization, it is crucial that city authorities incorporate into their strategies and policies the risks of social, technological and economic exclusion that can be compounded by the transformation of cities toward Smart Cities.

The first step city authorities should take is to identify and understand the specific needs and challenges facing different social groups in the city. Exclusion can range from access to technology to broader participation in social and economic life. Therefore, it is crucial that city policies be developed based on accurate data and analysis that identifies the most vulnerable groups to exclusion, such as the elderly, people with disabilities, ethnic minorities and low-income individuals. In this context, it is recommended that mechanisms be put in place to monitor and assess the situation of these groups in order to respond flexibly to their needs.

Another important aspect is the integration of the principles of inclusiveness and equality in the process of planning and implementing Smart City technologies. City authorities should ensure a situation, in which the city's technological development not only does not exacerbate existing social inequalities, but, on the contrary, actively supports the creation of equal access to city services and resources. This means that any decision on technological development must be made with a view to eliminating social and economic barriers that may limit access to new technologies. In practice, this means investing in technologies that are accessible and affordable to all residents, regardless of their age, economic status, education level or location. At the same time, special attention should be paid to the needs of marginalized groups that are often overlooked in traditional urban planning processes. An example is the development of smart transportation systems that take into account the needs of people with different types of disabilities, offering, for example, more accessible and easy-to-use user interfaces, or appropriate routes and stops.

Providing universal access to high-speed internet is also an important element of an inclusive Smart City. Network access has become an indispensable part of modern life, and its absence can exclude residents from using key digital services, such as online education, e-health or administrative services. Therefore, it is imperative that digital infrastructure development plans include the entire community, including rural and remote areas, where internet access may be limited.

Creating an inclusive and equality-oriented Smart City also requires active involvement of residents in decision-making processes. City authorities should promote dialog with diverse social groups to better understand their needs and expectations. Only in this way can the city's technological development be made to serve all residents, not just a select few. Creating spaces for active participation in public debates and community consultations, including for identified, potentially excluded groups, is key to making the Smart City truly responsive to the challenges and needs of the whole community. This approach builds a city of the future that is both modern and socially just. Solutions, such as e-democracy, citizen initiative platforms or community consultations, should be available and easily accessible, so that every resident can express their needs and concerns.

Smart City authorities should also place particular emphasis on education and awareness-raising among residents on the use of modern technologies. Education and training programmes should target all sections of society to ensure that every resident has the opportunity to participate fully in the digital society. In particular, initiatives that help older people and those with limited access to modern technology to acquire digital skills should be supported.

It is also important for Smart City authorities to systematically monitor and evaluate the effectiveness of the policies implemented to reduce exclusion. Regular evaluations and analyses will allow ongoing adaptation of strategies to changing social and technological conditions, as well as the identification of new areas, where interventions are needed. To this end, the introduction of standardized indicators and the involvement of local communities in evaluation processes are recommended, which will allow a better understanding of the real needs of residents and the adaptation of urban actions to their expectations.

The monograph makes an important contribution to science, especially in the context of the analysis and development of the Smart Cities concept. It provides valuable insights into the complexity of contemporary cities, which are becoming increasingly integrated with modern technologies. The monograph offers an interdisciplinary approach to the topic, combining perspectives from different fields, such as urban planning, sociology, information technology and public policy to provide a comprehensive understanding of Smart Cities.

An important contribution of the authors to the science is the presentation of the author's concept for assessing the inclusiveness of Smart Cities, combining indicator-based assessment and bottom-up measurement carried out on the basis of residents' opinions. A presentation of assessment directions and examples of indicators that can be used at a basic and advanced level, are components in this inclusiveness.

The monograph makes a significant contribution to research on inclusiveness and urban resilience to exclusion. The authors emphasize that the creation of Smart Cities should be based on an understanding of the specific needs of different social groups, including those most vulnerable to marginalization. The results of the analysis provide recommendations that can be used by policy-makers and urban planners to create more equitable and sustainable urban environments.

Another important aspect highlighted by the authors is the importance of social inclusion and civic participation in the process of Smart City development. The monograph analyses various models of urban governance that take into account the active involvement of citizens in the formation of urban policy. The work contributes to the knowledge of mechanisms that can increase public participation and strengthen democratic processes in cities of the future.

A valuable result of the conducted research is also the identification of phenomena that can negatively affect the risks of exclusion in Smart Cities, as well as those that aspire to become them and their connection to urban inclusions. Research in this area has made it possible to identify key factors that can negatively affect the process of social inclusion, that is, the inclusion of various social groups in full participation in urban life.

8.2 Summary, research limitations and directions for further research

The monograph provides a detailed analysis of the issue of social and digital exclusion in the context of the development of the Smart City concept. It describes in detail the various aspects and levels of exclusion that can occur in urban agglomerations striving to achieve the Smart City objectives and offers proposed solutions to increase inclusiveness and accessibility.

The authors have comprehensively reviewed the evolution of the Smart City concept from its origins, taking into account the successive stages of technological and social development. The content presents the successive developmental phases, from Smart City 1.0, which focused on automation and improving operational efficiency, to more advanced forms, such as Smart City 5.0, which emphasizes the integration of technical solutions with the needs of residents with a special focus on social and ethical aspects.

The paper presents issues related to the bright and dark sides of Smart Cities. For each of the 6 pillars of Smart Cities, an analysis was made of the positive and negative aspects of implementing the Smart City concept. For each area of Smart Cities, an analysis of the causes of exclusions was also made, along with ways to counteract these exclusions.

The monograph presents the role of rankings in the process of implementing the Smart Cities concept, which are used to evaluate and rank cities aspiring to be smart. A characteristic has been developed, including the most important rankings, together with their indicators, taking into account the exclusion dimensions.

In the preceding chapters of the monograph, rankings, such as IESE Cities in Motion Index, as well as the Smart City Index, have been used to present empirical findings in identifying risks associated with forms of exclusion in Smart Cities.

The results show that exclusion in the areas of sanitation and housing is significantly higher in cities from less developed regions, although the problem of housing availability also affects highly developed cities. In the area of education, on the other hand, cities from the top of the ranking offer better access to education. Moreover, the differences in education ratings between the best and worst cities are significant and illustrate the economic and social diversity of the cities. Acceptance of minorities is relatively evenly distributed, although cities from Australia and Europe received the highest ratings. Finally, corruption remains a significant problem, even in the most developed cities, although it is a much greater threat in cities at the bottom of the ranking. The analyses presented here suggest that economic and civilizational development promotes inclusiveness in Smart Cities, while low levels of these factors lead to an increased risk of social exclusion.

The monograph highlights significant challenges facing contemporary cities, such as the threat of digital, economic, educational and access to basic urban services exclusion. The authors analyse in detail a variety of indicators that identify the risks of exclusion in the context of different areas of city functioning, such as economy, mobility, environment, governance and social life. The analyses include a comprehensive approach to assessing the risk of exclusion in Smart Cities and, based on the analyses, strategies are proposed to manage these risks in order to increase the inclusiveness of Smart Cities by proposing the author's concept of inclusiveness assessment.

The work also draws attention to the need to include the voice of citizens in the Smart City policy-making process and the importance of transparency and accountability in urban governance to effectively address exclusion in its various forms.

This monograph is a valuable source of knowledge on the challenges and opportunities associated with the implementation of Smart City concepts. It presents a comprehensive picture of issues related to social and digital exclusion, while offering concrete recommendations for policy-makers and practitioners in the fields of urban planning and urban management that can contribute to building more sustainable and inclusive cities of the future.

The limitations of the research presented in the monograph primarily relate to the complexity and dynamism of the issues involved in the development of Smart Cities, which leads to the difficulty of accurately covering all aspects of the issue. One of the main limitations is the difficulty of taking into account the full diversity and specificity of urban contexts, which differ not only geographically, but also culturally, economically and socially. For this reason, some conclusions and recommendations may have limited universality and may not necessarily apply to all cities in an equally relevant way.

Another important limitation is the problem of rapid technological change. Technologies that currently are the subject of research and described in the context of Smart Cities may become obsolete in the coming years. This difficulty is also related to uncertainty about the future direction of technologies and their long-term impact on the operation of cities and the lives of residents. Nevertheless, since the monograph focuses more on presenting general concepts, rather than on the use of individual technologies, it seems that the analyses presented in it should not quickly become obsolete.

The availability and quality of data is also a limitation. Smart City studies often rely on data collected from different sources, which can be incomplete, inconsistent or difficult to compare. Indicators from different cities from different parts of the world may not be completely compatible.

Yet another limitation arises from the need to use certain simplifications and theoretical assumptions that may not fully reflect reality. By its nature, Smart City research must operate with models that, while helpful in understanding certain mechanisms, may overlook important factors that affect how cities operate in practice.

Finally, a limitation of the research is the lack of long-term data on the implementation of the Smart City concept. Due to its relatively new nature, many studies rely on short-term analyses, which can lead to overlooking the long-term consequences of implementing new technologies and urban strategies. The lack of a long-term perspective hinders a full understanding of the impact Smart Cities can have on society, the economy and the environment over the long term.

Future research on the people excluded in the context of Smart Cities should focus on a multidimensional analysis of the causes and consequences of exclusion in complex, technologically advanced urban environments. As cities become smarter, it is important to understand how modern technologies can both mitigate and exacerbate existing social inequalities.

One of the main directions for future research should be to analyse the impact of information and communication technologies on the accessibility and inclusiveness of urban services for the excluded. It is necessary to examine, how the development of technologies, such as the IoT, big data, AI and automation, affect different social groups, especially those most vulnerable to marginalization, such as the elderly, people with disabilities, migrants and low-income individuals. This research should consider both positive and negative aspects of the widespread use of modern technologies, highlighting potential barriers to accessing modern solutions, such as lack of digital skills or economic constraints.

Another important potential area of research is to examine how the infrastructure of urban Smart City systems can be designed and implemented in a way that minimizes exclusion. Research should be conducted on new urban and architectural models that take into account the needs of all residents, as well as analysing how existing infrastructure can be adapted to meet the needs of the excluded individuals. In this context, it is also important to understand how urban policies can support the development of accessible technologies and services that are affordable to people with varying levels of physical and mental ability.

Research should also focus on the social implications of the introduction of Smart City technologies. There is a need for an in-depth analysis of how the development of Smart Cities affects social relations and the sense of community among residents. It is important to study how different social groups perceive technological changes and what their concerns are about a future, in which technology plays an increasingly important role. This research should take into account psychological and sociological aspects, such as social alienation, changes in family structures and interpersonal relationships.

Another important direction for future research is to evaluate the effectiveness of public policies and initiatives to reduce exclusion in Smart Cities. It is worthwhile to study which strategies are most effective in addressing exclusion, as well as what tools and methods can be used to ensure that these strategies are effectively implemented in different cultural and economic contexts. In this regard, comparative studies that analyse the differences and similarities in approaches to exclusion in different cities around the world can be particularly valuable.

Bibliography

Batty, M. (2013). Big data, smart cities and city planning. *Dialogues in Human Geography*, 3(3), 274–279.

Calvillo, C. F., Sánchez-Miralles, A., Villar, J. (2016). Energy management and planning in smart cities. *Renewable and Sustainable Energy Reviews*, 55, 273–287.

Cardullo, P., Kitchin, R. (2019). Smart urbanism and smart citizenship: The neoliberal logic of 'citizen-focused' smart cities in Europe. *Environment and Planning C: Politics and Space*, 37(5), 813–830.

Hollands, R. G. (2008). Will the real smart city please stand up? Intelligent, progressive or entrepreneurial? *City*, 12(3), 303–320.

Kitchin, R. (2015). Making sense of smart cities: Addressing present shortcomings. *Cambridge Journal of Regions, Economy and Society*, 8(1), 131–136.

Kitchin, R. (2016). The ethics of smart cities and urban science. *Philosophical Transactions of the Royal Society A: Mathematical, Physical and Engineering Sciences*, 374(2083), 1–13.

Shelton, T., Lodato, T. (2019). Actually existing smart citizens: Expertise and (non)participation in the making of the smart city. *City*, 23(1), 35–52.

Townsend, A. (2013). *Smart cities: Big data, civic hackers, and the quest for a new utopia*. New York: WW Norton.

van Deursen, A. J. A. M., van Dijk, J. A. G .M. (2014). The digital divide shifts to differences in usage. *New Media and Society*, 16(3), 507–526.

Vanolo, A. (2014). Smart mentality: The smart city as disciplinary strategy. *Urban Studies*, 51(5), 883–898.

Conclusions

The considerations in this monograph are entirely devoted to a rather narrow research problem of exclusion in Smart Cities. Four key perspectives were used to review this problem. Firstly, those exclusions that have already been present in urban space for many years, and which may include, for example, corruption or crime. Secondly, new exclusions, generated by Smart Cities, were also included in the considerations, among which digital exclusion can certainly be placed. Thirdly, civilizational and economic exclusions concerning differences between cities were reported. And fourthly, contemporary socio-economic trends were identified that may exacerbate existing exclusions in the future or generate new types of exclusion.

The authors of the monograph also attempted to develop a general concept for assessing the risk of exclusion, taking into account indicator and civic assessment. They also proposed an outline of a strategy for strengthening the resilience of cities to exclusions and enhancing their inclusiveness. This strategy includes a case-by-case approach to infrastructural and social exclusions. Eliminating exclusions from the first group requires technological and design as well as financial measures. The mitigation of exclusions from the second group should be based on diversity management solutions and their adaptation to the specifics of the local community.

The elements that distinguish the monograph from other Smart City publications are as follows:

- focusing exclusively on the issue of exclusion;
- detailed analysis of the types of exclusions,

DOI: 10.4324/9781003499992-10

- identification of the scope of measuring inclusiveness and exclusion in selected Smart City evaluation studies;
- a preliminary proposal for assessing exclusion risks;
- development of assumptions for strategies to strengthen the resilience of cities to exclusions and the development of inclusiveness; and
- identification of potential sources of intensification of exclusion threats in the future.

Index

Note: Tables and figures are indicated by page numbers in **bold** type and *italic* type respectively.

For Product Safety Concerns and Information please contact our EU representative GPSR@taylorandfrancis.com
Taylor & Francis Verlag GmbH, Kaufingerstraße 24, 80331 München, Germany

www.ingramcontent.com/pod-product-compliance
Lightning Source LLC
LaVergne TN
LVHW020712110826
845149LV00012B/2233
* 9 7 8 1 0 3 2 8 1 4 6 5 0 *